Intervening Spaces

At the Interface/Probing the Boundaries

Series Editor

Rob Fisher (*Oxford, United Kingdom*)

VOLUME 107

The titles published in this series are listed at *brill.com/aipb*

Intervening Spaces

Respatialisation and the Body

Edited by

Nycole Prowse

BRILL
RODOPI

LEIDEN | BOSTON

Many thanks to Anthi Kosma, one of the contributors to this volume, for creating the cover illustration and this accompanying poem, 'Beyond the Image':

Memories reside
beyond the Image
beyond the traces.

Images of memories
become then, memories of traces.
Memories of bodies tracing, writing, making.

The Library of Congress Cataloging-in-Publication Data is available online at http://catalog.loc.gov
LC record available at http://lccn.loc.gov/2018942457

Typeface for the Latin, Greek, and Cyrillic scripts: "Brill". See and download: brill.com/brill-typeface.

ISSN 1570-7113
ISBN 978-90-04-36299-4 (paperback)
ISBN 978-90-04-36552-0 (e-book)

This book is printed on acid-free paper and produced in a sustainable manner.

Contents

Notes on Contributors

Nycole Prowse
is a writer, holds a PhD in literature and teaches into literature at the University of the Sunshine Coast Australia and is the Principal of Peripheral Arts. Her current research focuses on literary representations of women, addiction and drug use. Her broader research base includes historical approaches to feminist discourse and philosophy as well as postmodern feminist concerns of culture and difference.

Burcu Baykan
holds a PhD in Digital Arts from Trinity College Dublin and is an assistant professor of visual communication design at Bilkent University, Turkey. She has published articles and chapters in edited collections on contemporary art.

Michelle Collins
completed a Masters in Traditional Arts at the University of South East Norway, Rauland in 2014 and graduated from the University of Manchester in 2016 with a Masters in Social Anthropology. With a background in sociology, music and psychotherapy, this author's current interests include anthropology of the sound, ritual, sound studies, modern mythologies, and themes of magic, beauty and shame.

Katsuhiko Muramoto
is Associate Professor at the Department of Architecture, Pennsylvania State. His professional office experience includes positions with Peter Eisenman, Richard Meier and Ohbayashi, and he was co-founder of Cover LLC, a design/build architectural office in Philadelphia. His research agenda encompasses architectural design, theory of twentieth-century Japanese architecture, and architectural representation and digital mediation. He is a co-editor of the book, *Energy Accounts* (2016).

Shelley Smith
holds a BA AID in Interior Design from Ryerson University in Toronto, a BA and MA in architecture, and a PhD in urban design from the Aarhus School of Architecture. She is an associate professor at Aalborg University, Denmark, where she teaches urban design. In her research, she focuses on informal use of contemporary urban space as a function of urban space design with a particular interest in parkour.

Nicolai Steinø
holds a BA and MA in architecture, and a PhD in urban design from the Aarhus School of Architecture, and a Graduate Diploma in Business Administration from Aalborg University. He is an associate professor at Aalborg University, Denmark, where he teaches form-making and urban design. In his research, he focuses on form-making, and parametric urban design as a tool for participatory urban design processes.

Ana Carolina Lima e Ferreira
was born in Brasilia, Brazil. She graduated in Fine Arts. In Germany she studied her Master of Fine Arts and her PhD titled 'The Theatre of the Oppressed as Media of Public Arts' at the Bauhaus-Universität Weimar, Germany.

İklim Topaloğlu Topaloğlu
is a Master Architect in the Department of Urban Design and Urban Aesthetics at İzmir Metropolitan Municipality. She completed Architectural Design Master Programme in Istanbul Technical University. Her research interests include architectural installation studies and their effects on public spaces.

Pelin Dursun Çebi
received her PhD degree in 2002 from ITU Faculty of Architecture. Her research interests focus on architectural design, design education, body and space relationships, architectural morphology and space syntax. Assoc. Prof. Dr Dursun is teaching in ITU Faculty of Architecture at undergraduate and graduate level.

Katerina Mojanchevska
is a PhD student at the International Institute of Social Studies (ISS), part of the Erasmus University Rotterdam, The Netherlands. She has a professional experience in the civil society sector in Macedonia working on cultural policy and urban development projects. Her research interests encompass intersection among identity, public space, inter-culturality and urban policy.

Clementine Monro
is Programme Leader for BA (Hons) Photography and Senior Lecturer, in The Lincoln School of Film and Media, at the University of Lincoln, UK. Her research and practice explores the relationship between photography and absence.

Christobel Kelly

MVA (2015), UniSA, lectures in Art History at Adelaide College of the Arts. She is a Fellow of the Ballinglen Art Foundation, Republic of Ireland. *Voyages and Fictions* is published in, Imprint, (Print Council of Australia, 2017). Her art practice includes printmaking, painting and sculpture.

Anthi Kosma

(Technical University of Madrid) received her diploma of architect – engineer from DUTH (Greece, 2005). She holds a PhD and DEA from ETSA of Madrid, UPM (Spain, 2014) with a scholarship from IKY and Triantafyllidis foundation. You can find written, drawn and other traces of her at: http://imprografika.wordpress.com/

Introduction

Nycole Prowse

Intervening Spaces is a dialogic-themed volume that brings together a cross-disciplinary examination of the interconnectedness between bodies, time and space. The title is suggestive of the oscillating and at times political impact that occurs when bodies and space engage in non-conventional ways. Bodies *intervene* with space, creating place. Likewise space can *intervene* – re-orientating bodies. It is this 'oscillating impact' that reconceptualises notions of the subject-body and the space it inhabits. And it is this juncture that encapsulates what is meant by the notion 'respatialisation.' Such respatialisation does not occur in a temporal vacuum nor does a sense of ontological becoming need to span lengthy time periods. The moment can be more significant than millennia in producing new ways to see corporeal connections with space.

Respatialisation is discussed across disciplines via the collapsing of dichotomies such as centre/margin, real/imagined, life/death, absence/presence, and inside/outside. These temporal and spatial dichotomies are dissolved, disrupted, interrupted and erupted via interventions – revealing the potential and possibilities of respatialisation. Interrupting and intervening spaces as examined in this volume helps to generate, in Elizabeth Grosz' words, 'new perspectives, new bodies, new ways of inhabiting'.[1]

The volume explores the key theoretical and thematic threads that were revealed at the Fourth Inter-Disciplinary.Net 2014 'Time, Space and the Body' conference in Oxford, UK. In particular, the volume outlines a clear and recurring connection between disciplines based on the notion of respatialisation and its potentiality for the subject-body. The anthology draws upon the body's interpolation with space in literary spheres, digital worlds and urban built environments. It also examines spaces in-between, liminal worlds where the subject-body slips past conventional notions of subjectivity. The interdisciplinarity of the volume mirrors the permeability of the body and space, resultant of the impact of respatialisation. This spatio-temporal interconnectivity of the body across disciplines informs harmonious ways to live, to plan and to play – to intervene with space and each other.

This volume engages in an already established and rich theoretical analysis of intervening spaces. As with Grosz' diachronic and dialogic analysis of theoretical approaches to time, space and the body in her seminal texts

1 See Elizabeth Grosz, *Space, Time and Perversion* (New York: Routledge 1995).

 | DOI 10.1163/9789004365520_002

Volatile Bodies (1994) and *Space, Time and Perversion* (1995), this anthology uses a number of similar theoretical vantage points to understand the different spatial, temporal and corporeal interventions that evoke an understanding of respatialisation. In particular the volume draws on theorists as diverse as Michel Foucault, Giles Deleuze, Félix Guattari, Martin Heidegger, Tetsuro Watsuji Jacques Lacan, Maurice Merleau-Ponty, Henri Lefebvre, Judith Butler, Elizabeth Grosz and Margrit Shildrick, among others. Each theoretical vantage point adds to the volume's examination and reconceptualisation of time, space and the body. In this way the anthology reframes conventional notions of the body to observe the interconnectedness of the body in time and space revealing new corporeal possibilities and potential.

As a dialogic-themed volume, chapters are organised into three broad thematic sections.

Section One contains chapters that examine respatialisation via interventions upon the body that collapse spatial, temporal and corporeal borderlines. For example Nycole Prowse examines the way drug imagery evokes leakages and slippages across time, space and the body enabling a re-evaluation of corporeal possibilities and potential. Comparatively, while Prowse's focus is on drugged interventions on the body, in the subsequent chapter Burcu Baykan poses similar questions about altered bodies (via surgical intervention/performance) via a similar theoretical vantage point. Corporeal interventions are also explored in Michelle Collins' examination of the transformation of the body via keening, aligning with Prowse and Baykan's assertions of the potential of bodily transformation that transcend conventional notions of space. In the final chapter of the first section, Katsuhiko Muramoto extends the understanding of the consequences of re-conceptualising notions of the body and the body schema to reveal new understanding of the spatio-corporeal relationships and possibilities.

Like Prowse, Baykan draws on Deleuze and Guattari's conception of the 'intermezzo'; the rhizomatic space of in-betweenness and Grosz's formulations on the relation of body to space, interiority and the exterior. Examining the corporeal possibilities explored by Orlan's performative and digital experiments, Baykan draws on similar conceptual tools used by Prowse as well as the collapsing of bodily-spatial dichotomies in Collin's chapter to argue that mutable states of being occupy an in-between milieu that characterizes the intermediary rhizomatic space. Baykan's analysis further connects with Muramoto in considering practices and corporeal reconceptualisation that enables proliferating connections and fosters non-dualistic forms of bodily space that is no longer contingent on binary divisions.

Collins' chapter also resonates with Prowse and Baykan in the way it considers spatio-corporeal transformation through use of the voice and body within constructed spaces. It further draws upon this section's focus on bodily interventions and re-conceptualisations in creating the liminal 'betwixt and between' state that transgresses thresholds of bodily interiority-exteriority in transforming communal and cultural space.

Muramoto's approach corresponds with the other chapters in this section in the way it reframes an understanding of the interconnectedness and interdependedness between technology, environment and societal values. Through a re-conceptualisation of the body and space via the extension of the body schema through architectural design, Muramoto's examination exemplifies the transformative potential of respatialisation present in the other chapters. Evoking similar theoretical understanding of the concept of 'co-existence' that elucidates reciprocity (Heidegger, Watsuji and Merleau-Ponty) between the individual, collective, built environment and nature, Muramoto connects with the other chapters' reframing of the body to collapse and transcend conventional and restrictive spatio-temporal and corporeal borderlines.

Section Two focuses on bodily interventions on material space. Using the practice of parkour as a vehicle for thought and exemplification, Shelley Smith and Nicolai Steinø align with Ana Carolina Lima e Ferreira's subsequent chapter on theatrical performance – the performative body and its influence and intervention on public space. Design and performance interventions on public space is further explored by İklim Topaloğlu and Pelin Dursun Çebi in their analysis of public art installations. Smith's and Steinø's point of departure in alternative urban practices and the role of architecture and design in materialising space also links with Katerina Mojanchevska's examination of the impact of citizens on respatialising public spaces. Smith's and Steinø's paper incorporates this section's examination of the potential of bodily interventions on urban spaces aiming for plurality and ambivalence, rather than uniformity and unambiguity.

Like Smith, Steinø, Topaloğlu and Dursun Çebi, Lima e Ferreira analyses the implementation of projects and active participation of citizens to politicise space. As with Mojanchevska's assertions on the potential of the public voice on urban spaces, Lima e Ferreira alludes to Boal's theatrical interventions and group actions as democratising public spaces as cultural products as possibilities to activate, strengthen and empower citizenship. In line with the spatial transformations of other chapters in this section, Lima e Ferreira explores the way that implementation of community projects and active participation creates change via urban intervention.

As with Lima e Ferreira's work, the aim of Topaloğlu's and Dursun Çebi's study is to focus on participant and audience-oriented art and architectural installation works and to discuss their contributions to the urban space. Akin to Smith's, Steinø's and Mojanchevska's assertions, Topaloğlu and Dursun Çebi examine spatial potential in the way bodies transform public space. Here it is believed that decoding design enhances spatial awareness and helps both designers and architects to create innovative concepts for re-definition of public space.

In line with the research focus of this section, Mojanchevska advocates participatory approaches that operate beyond representational democracy and voicing the interests and perspectives of different social groups in respective bodies and spaces. Mojanchevska asks if and in which ways do self-organising citizens' groups install spaces of participation that shift the power balance in urban and social making of cities and if the control over public planning process is real or symbolic. Each chapter in this section examines the transformative impact of corporeal interventions and reconceptualisations on public spaces.

Finally, *Section Three* examines artistic interventions on space evoking a sense of spatio-temporal-corporeal *becoming*. It is a section that incorporates visual artists and their practice led research that integrates theory 'through immersion of the body into space, locating absences and traces…movement of bodies through space.'[2] Clementine Monro incorporates the concept of (corporeal) absence in creating presence in space that is examined in each of the subsequent chapters of this section. The word presence, which means existence in place, is in direct contrast to absence as meaning the non-existence or lack of, but as Monro suggests, the two are intrinsically connected. There is presence within absence and this can be felt through the spatial experience of the ruin. As with Christobel Kelly's phenomenological practice-led exploration of the 'lived landscape,' Monro explores the capacity of photography to communicate time, duration and the ephemeral. Anthi Kosma's chapter on the collapsing of corporeal temporal and spatial boundaries through her analysis of tracing resonates with Monro's assertions on the way photography mediates a visual trace and expression of the felt experience of absence.

As with Monro's assertions of the corporeal impact on space, Kelly's research takes the form of a bodily response to the physical landscape. It deals with ways in which the rhythms and modes of walking lead to a special way of thinking that sits outside our *modus vivendi*. It also investigates how familiar

2 Christobel Kelly, 'Colonial and Gothic Dissonance: Tracing Desire Lines through Postcolonial Landscape,' in this volume.

landscape, which is hidden in plain sight by quotidian experience, can become unfamiliar through contrasting times of day and shifting weather. In comparison to Kosma's understanding of tracing in reconceptualising the body in space, Kelly proposes that layers of meaning to do with time, memory, space and the body can only become available when sensory perceptions are heightened through the act of walking.

Kosma incorporates a number of aspects of each of the preceding chapters by revealing the potential of 'tracing' to understand the body's impact on space as well as the implication this action has on a sense of agency and becoming. As Kosma suggests, drawing is described as an action of coming-into-presence which means presence before signification and also as a non-representative art where the form is not the traced figures, but the body as a form by itself, the body as generator of space. Additionally, the experience of space is presented in its formation and not as a mere object or as a reflection about a finished object. If contemporary drawing can be defined as a 'description without a place,'[3] Kosma looks forward by describing events where space is presented as an out of place experience.

The rationale for this anthology is further evidenced in its continuation of the engaged and, surprisingly, harmonious (considering the diverse and interdisciplinary nature of the conference) discussions that began with the fourth 'Time, Space and the Body' conference. Indeed, the interconnectedness between the 'body' of delegates and the space created at the three-day conference reflects the ethos and focus of this volume. The authors span a wide range of disciplines and countries of focus. Included in this volume are studies related to Australia, Ireland, Denmark, Greece, Turkey, Macedonia, Spain, Canada, Brazil, Japan, Germany, England, The Netherlands and the United States of America. It is an important contribution that will be of interest to academics and students, especially in the fields of Sociology, Literature, Performance Arts, Visual Arts, Geography, Architecture, Urban Design and History.

3 Alain Badiou, 'Drawing.' Lacanian ink 28 (2006): 42–48. September 12, 2011. v viewed on 25 January 2015, http://lacan.com/symptom12/?p=65.

PART 1

Reconceptualising the Body: Interventions on the Body that Enact Respatialisation

CHAPTER 1

Possibilities from the Peripheries: Time, Space and the Body in the Australian Drug Writing of Helen Garner and Luke Davies

Nycole Prowse

Abstract

This section begins with an analysis of the leaky representations of time, space and the body in drug literature. It is selected as the opening chapter of this section because of its theoretical approach, as well as the thematic focus of the way in which interventions upon the body (via the drug trope) are reflective of the way the body is inhabited by and inhabits space. For this examination, this chapter utilises Elizabeth Grosz's understanding of how bodies live and are positioned as spatio-temporal beings. Grosz suggests that '...in order to reconceive bodies, and to understand the kinds of active interrelations possible between (lived) representations of the body and (theoretical) representations of space and time, the bodies of each sex need to be accorded the possibility of a different space-time framework'. This chapter proposes that drug literature provides a platform where alternative corporeal possibilities can be played out. In the literary sphere the drug trope reframes spatial and temporal regulatory notions of the body. The drug metaphor disrupts temporal linearity through the reconfiguration of junk time. Likewise, landscapes, cityscapes and a sense of place are re-imagined in fluid, drugged dreamscapes.

1 Introduction

> Women are practised on the peripheries ... Our memories, our stories, like the ways we live, are formed in movement between inner and outer, past and future, centre and margin, between the physical environment and the social world. We shape our cities, and re-shape them from the edge, we always have; just as our cities shape us.[1]

1 Drusilla Modjeska ed., *Inner Cities: Australian Women's Memory of Place* (Ringwood: Penguin Books, 1989), 2.

 | DOI 10.1163/9789004365520_003

Australian realist drug novels *Monkey Grip* (1977)[2] by Helen Garner and *Candy* (1997)[3] by Luke Davies use the drug trope to reshape the concepts of time, place and the body in an organic way and with much less apocalyptic outcomes than other drug writing such as that of W.S. Burroughs. Garner and Davies employ pastoral imagery to portray junk bodies inhabiting and (in) habited by inner-city urban space, radicalising traditional notions of subjectivity and a sense of place in realist Australian fiction. Contrary to the dystopian elements of American and British postmodern novels of the 1960s, *Monkey Grip* and *Candy* reframe conventional pastoral narratives. Garner and Davies use natural imagery to re-evaluate the urban corporeal experience. Resonating with Burcu Baykan's[4] Deleuzian analysis of Orlan's subversion of spatial dualities in this volume, the drug trope in Garner's and Davies' novels intensifies and collapses spatial divisions between the rural and urban, providing a platform where alternative possibilities for the subject-body can be examined and played out. The drug metaphor also disrupts temporal linearity through the reconfiguration of 'junk time.' Likewise, landscapes, cityscapes and a sense of place are re-imagined by the fluidity of the drug trope. While Garner's and Davies' novels both use the drug trope to reframe spatial and temporal regulatory notions of the body, they do so differently and with contrasting outcomes. The gendered differences found in the representation of drugged bodies in each of their novels illuminate the corresponding corporeal connections or disconnections with space and time.

The use of the drug trope in *Candy* brings to attention the displacement of male bodies that do not conform to the regulated notions of time and space. Paradoxically, however, Davies' novel situates the male protagonist as disconnected from nature, perpetuating conventional conceptions of the body as separate from the space it inhabits. As Katsuhiko Muramoto[5] alerts us to in this volume, the separation between man and nature, 'a one-way mode of causation' stems from the 'Cartesian paradigm of subject-object duality.' This is proliferated in Kantian notions of self. Kant argues in *Critique of Pure Reason* (1787) for an objectification of the external world because we shape our experience of things, the external world, through the filter of our mind and

2 Original publication date.

3 Original publication date.

4 Burcu Baykan, 'Respatialising the Body: The Ontological In-between Subject in Orlan's Body of Work,' in this volume.

5 Katsuhiko Muramoto, 'In Search of a Relational Body: Reflections on Sustainable Architecture,' in this volume.

in so doing 'objects must conform to our cognition'.[6] In a Kantian sense the (male) subject of *Candy* observes nature as external to the body, however, Davies' portrayal of nature is foreboding, leaving the male protagonist stripped of agency. Davies' novel also accentuates the way women's bodies are represented as doubly marginalised or 'already' displaced within patriarchal regulatory systems. The drug trope in *Candy* enables a tragic exposure of displacement and marginalisation of bodies constrained by conventional notions of time, space and bodies.

On the other hand, the re-envisioning of the temporal and spatial positionality of women in Garner's novel reveals the interconnection of time, space and the body. In line with Baykan's[7] discussion of space and the body in Orlan's work, the drug trope is used in *Monkey Grip* to challenge conventional/patriarchal representations of the relationship between the subject and space. While Baykan draws on Deleuze to explicate Orlan's collapsing of spatial boundaries, the opiated dreamscapes depicted in Garner's novel blend the urban and the rural in a Spinozan sense to entwine the subject organically within the space he/she inhabits. Spinoza's positioning of the subject-body as a piece of the natural world reframes the significance of the external upon the individual by challenging binary notions of space as arbitrary. Spinoza postulates that: 'the first element which constitutes the actual being of the human mind, is the idea of some particular thing actually existing'.[8] Hayter places Spinoza, 'fairly and squarely within the "It is" category of philosophers'.[9] The use of the drug trope to reveal the spatio-corporeal interconnectedness is akin to Muramoto's[10] use of Merleau-Ponty's 'corporeal reflexivity' and Watsuji's exposition of 'between-ness' in his analysis of 'interdependence between human and nature' in Japanese architecture. Like Muramoto's exposition of Watsuji's 'symbiosis of inside and outside'[11] the protagonist/narrator of *Monkey Grip* observes and experiences nature from within nature. It is a position that subversively challenges notions of the marginalised. Furthermore it is this reconceptualisation of space via the drug trope that connects with Michelle Collin's[12] assertions

6 Immanuel Kant, 'Critique of Pure Reason,' *Philosophy on the EServer* (Ames: Iowa State University, 2014), np, viewed 25 July 2014, http://www.eserver.org.

7 Baykan, in this volume.

8 Alethea Hayter, *Opium and the Romantic Imagination* (Berkeley: University of California Press, 1968), 122.

9 Ibid., 122.

10 Muramoto, in this volume.

11 Muramoto, in this volume.

12 Michelle Collins, 'Corporeal Interventions and the Contemporary Sounds of Keening,' in this volume.

in this volume of liminality as an empowering position of habitation on the edge. Garner's novel comes from the dangerous borderlines, a site of feminist political engagement. Garner's use of the drug trope is suggestive of the way in which the female subject belongs to space, is emplaced in temporal and spatial surrounds. This is evidenced in the way Garner uses the drug trope to re-envision time and space from a female corporeal position in *Monkey Grip* – one that opens up the potential of resistance, empowerment and agency – from the peripheries.

2 Reconceptualising Space, Time and the Body

A comparative textual analysis of Garner's *Monkey Grip* and Davies' *Candy* provides a site to canvas Grosz's[13] assertions on the implication of gender when reframing the spatio-temporal positionality of bodies. Grosz's seminal text, *Space, Time and Perversion*,[14] initiated a postmodern feminist understanding of how bodies live and are positioned as spatio-temporal beings. Grosz suggests that

> ... in order to reconceive bodies, and to understand the kinds of active interrelations possible between (lived) representations of the body and (theoretical) representations of space and time, the bodies of each sex need to be accorded the possibility of a different space-time framework.[15]

A postmodern feminist examination of Garner's and Davies' novels can test the arbitrary nature of space, time and bodies as cultural, and more specifically, patriarchal constructs. Such reconceptualisation destabilises the constraints of spatio-temporal constructs upon the (female) subject. As Irigaray suggests

> The question of the dwelling, of where and how to live, is ... a crucial one both in the production of the male domination of women's bodies, and in women's struggles to acquire an autonomous space they can occupy, and live in as women.[16]

13 Grosz, *Space, Time and Perversion.*
14 Ibid.
15 Ibid., 100.
16 Luce Irigaray, *Speculum of the Other Woman*, trans. Gillian C. Gill (Ithaca: Cornell University Press, 1985), 143–144.

More specifically it is the use of the urban space in Garner's and Davies' novels that will be examined for their potential to re-conceptualise bodies, sexed bodies that inhabit that space. Grosz suggests that the city is the site where the body can be re-explored to see how the city impacts on how the subject sees others and our alignment with others:

> The city is one of the crucial factors in the social production of (sexed) corporeality: the built environment provides the context and coordinates for contemporary forms of body. The city provides the order and organization that automatically links otherwise unrelated bodies: it is the condition and milieu in which corporeality is socially, sexually, and discursively produced.[17]

The use of the drug trope in Garner's and Davies' urban novels will be examined for the way in which it can elucidate the connections/leakages that occur between bodies and their urban habitation. The destabilisation of corporeal borderlines via the drug metaphor aligns with Grosz's assertions of corporeal 'pliability' 'stretchability' within the city-sphere.[18] The city, Grosz suggests, 'seeps' into the body and 'affect[s] all the other elements that go into the constitution of corporeality'.[19] Grosz also discusses the way in which the body 'reinscribes' the city.[20] The city space procures 'the body's cultural saturation' transformed, re-examined, and re-explored via representational systems:

> ...the body (as cultural product) transforms, reinscribes the urban landscape according to its changing (demographic) needs, extending the limits of the city ever towards the countryside that borders it.[21]

The city, like the literary space, can be seen as a platform on which to formulate questions, exposing the displacement of the subject. It can also be a site to find answers, a place of corporeal alternatives – 'emplacing' the subject. If the city is, as Grosz suggests, 'the most immediate locus for the production and circulation of power,'[22] the literary city can be seen as a site to re-produce bodies to

17 Grosz, *Space, Time and Perversion*, 104.
18 Ibid., 108.
19 Ibid.
20 Ibid.
21 Ibid., 108–109.
22 Ibid., 109.

become subjects and space to become place. Sara Ahmed's assertions on the body in space also pose a sense of possibility via re-orientating bodies in space that opens up alternative ways of being.[23] For Ahmed

> ... orientation is a matter of how we reside, or how we clear space that is familiar .... Even when orientations seem to be about which way we are facing in the present, they also point us toward the future. The hope of changing directions is that we don't always know where some paths may take us: risking departure from the straight and narrow makes new futures possible, which might involve going astray, getting lost, or even becoming queer.[24]

This 'queering' of space, or what Ahmed refers to as a 'disorientation' has resonance with the way the drug trope impacts space and the body in Garner's and Davies' novels.[25] The positive, utopian connotations of Ahmed's re or dis-orientation of the subject and its habitation, however, coalesces more so with Garner's celebratory use of the drug trope.

The urban realism of Garner's and Davies' novels is mimetic of societal functioning. Their use of the drug trope hyperbolises – magnifies societal dysfunction creating a hyper-realism. Watson and Gibson discuss Benjamin's critique of the modern city as a 'narcotic dream – a "phantasmagoria", the dream world of the urban spectacle' but one that perpetuates a 'utopian longing for something other than this urban labyrinth'.[26] For Benjamin the urban scene 'comes to represent utopia and dystopia simultaneously'.[27] It is this disjuncture that may explain the urban space represented in both Garner's and Davies' drug novels. Garner's and Davies' use of the drug metaphor will be explored for its illumination of the permeability of the subject and space. Space, time and body are hyperbolised via the drug trope magnifying and enabling the interconnectivity of each and the diversity and difference.

Like the oscillation of impact of space and the subject, time is also ambivalent in the drugged urban space of Garner's and Davies' novels again aligning

23 Sara Ahmed, *Queer Phenomenology: Orientations, Objects, Others* (Durham: Duke University Press, 2006a).

24 Ibid., 20.

25 Ibid.

26 Sophie Watson and Katherine Gibson eds., *Postmodern Cities and Spaces* (Oxford: Blackwell, 1995), 73–74.

27 Ibid., 74.

with Grosz's assertions on subjectivity. For Grosz 'time cannot be viewed directly.'[28] Grosz suggests

> Time is not merely the attribute of a subject, imposed by us on the world: it is a condition of what is living, of matter, of the real, of the universe itself. It is what the universe imposes on us rather than we on it; it is what we find ourselves immersed in, given, as impinging and as enabling as our spatiality.[29]

The subject-body is at time's disposal no matter how much we try to control or measure it. Grosz suggests that time is 'perhaps the most enigmatic, the most paradoxical, elusive, and "unreal" of any form of material existence'.[30] The ontological possibilities of time as continual becoming are reflected in drug literature. The drug trope reflects the paradoxes of time and the drugged protagonist re-enacts this. Via the drug trope we are 'jarred out of our immersion in its continuity, when something untimely disrupts our expectations'.[31] The drug trope disrupts conventional expectations of time allowing us to think it, to be aware of it and to analyse it. Drugs are temporally subversive – they magnify, make clear, and centre time so as to examine its impact, our belonging to it. It is the drugged 'exploration of those nicks, disruptions or upheavals – events that disrupt our immersion in and provoke our conceptualization of temporal continuity',[32] where political engagement can take place. The drugs provide a distortion of the conventional conceptions of time and held in literary narratives, these moments of time can be refocused. Grosz' understanding of time as paradoxical – and the impact that this has on the subject – explains junk's destabilisation of traditional concepts of time – evoked by the drug trope in literature.

Through the centralising of the drugged protagonist, the reconceptualisation of time and space in Garner's and Davies' drug writing exemplifies Stallybrass and White's re-positioning of the 'perverse,' the 'scatological' and the 'low' in 'expos[ing] the hidden truth' about society.[33] Alluding to Bakhtin's

28 Elizabeth Grosz, *The Nick of Time: Politics, Evolution and the Untimely* (Crows Nest: Allen & Unwin, 2004), 5.

29 Ibid., 4–5.

30 Ibid., 4.

31 Ibid., 5.

32 Ibid.

33 Peter Stallybrass and Allon White, *The Politics and Poetics of Transgression* (London: Methuen, 1986), 20.

assertions on the carnivalesque, Stallybrass and White suggest that '[w]hat is socially peripheral is often symbolically central ... out of all proportion to their actual social importance'.[34] Garner's and Davies' drug novels will be explored for their inversion of spatial and temporal power structures, destabilisation of notions of subjectivity and creation of a 'perverse' sense of self from the peripheries of the symbolic order.

3 Down under and Out of Time: Australian Drug Literature

Australian drug literature, particularly the realist novels of Garner and Davies are especially appropriate texts to analyse the power of peripheries for their protracted sense of spatial and temporal distortion. Australian Literature has been and perhaps still is seen to be located on the peripheries. In Pablo Armellino's analysis of the 'ob-scene' spaces in Australian narrative, he suggests that Australia's situation as a region 'down under' is a result of the 'bourgeois unconscious' historically displaced in an 'ob-scene' setting: 'being at the antipodes of the globe and with an apparently bare and void landscape'.[35] Australia's geographical and temporal global positioning also impacts upon the subject-body. According to Graeme Turner literary representations of 'the Australian identity' evoke a sense of self as imprisoned, exiled and divorced and 'as such, they seem suggestive and productive metaphors for Australian existence'.[36] Garner's and Davies' post 1970s novels, reconceptualise conservative literary representations of the Australian national identity, by reframing traditional notions of the bush and the city and providing alternatives to being in place.

4 Mapping the 'Constricted Corridor':[37] Space, Time and the Body in Luke Davies' *Candy*

Garner's and Davies' consistent use of natural imagery is antithetical to much contemporary grunge, drug literature, such as Irvine Welsh's *Trainspotting* (1993) or Ann Marlowe's *How to Stop Time: Heroin from A to Z* (2002), and is

34 Ibid.

35 Pablo Armellino, *Ob-Scene Spaces in Australian Narrative: An Account of the Socio-Topographic Construction of Space in Australian Literature* (Stuttgart: Ibidem-Verlag, 2009), 195.

36 Graeme Turner, *National Fictions: Literature, Film and the Construction of Australian Narrative* (Sydney: Allen & Unwin, 1986), 84.

37 Barry Lopez, Front Cover of *Candy* by Luke Davies (Crows Nest: Allen and Unwin, 2006).

perhaps indicative of a distinctly Australian sense of the urban. Despite this common stance, the key differences between the two novels actually lie in this juxtaposition of the city space and nature. Although both novels overlay the inner city urban space with natural imagery – the opiated cities are produced differently, providing contrasting outcomes for the subject. The differences can be explained by what Judith Wright has called the 'double aspect' of the Australian version of nature – its dualistic ability to simultaneously represent both the 'reality of newness and freedom' and the 'reality of exile.'[38] As will be evidenced later in this chapter, Garner uses natural imagery in the urban setting of *Monkey Grip* to signify an alternative urbanised pastoral ideal – a 'newness and freedom.'[39] In this way Garner uses the drug trope to debunk the pastoral myth of the Australian fictionalised bush setting by shifting the rustic serenity of the pastoral to portray the urban sphere. Davies' images of nature draw on the other side of Wright's double aspect – natural disasters, droughts and flash floods culminate in a harsh environment that disempowers the subject.

Candy aligns with traditional stereotypes of Australian identity in exile. As with other male drug writers, like Burroughs and Samuel Taylor Coleridge, Davies perpetuates a Kantian estrangement of the subject (I Am) from nature. Davies portrays space as something external. Nature is initially idealised in a pseudo-utopian sense as a place to 'withdraw' and be free of addiction. This is exemplified in the following passage from *Candy*:

> ... in a month – we gave ourselves a time limit – we would go down to Gippsland to dry out. We had a couple of friends who lived there, a couple of hours out of Melbourne, in rich green dairy country .... They'd settled into a pretty stable relationship...and a nice life down there, smoking buds and gardening and painting. Their offer had always been there – if our intentions were good, we were welcome to come down and make use of their spare room and good food and quiet lifestyle.[40]

However, as discussed below, nature is overridingly represented in Davies' novel as brutal and harsh through the imagery of natural disasters. The urban is portrayed in *Candy* as ordered and mechanical with 'streets filled with the usual Saturday morning buzz of activity'[41] and yet also as a city of foreboding

38 Turner, *National Fictions: Literature, Film and the Construction of Australian Narrative*, 25.

39 Ibid.

40 Luke Davies, *Candy* (Crows Nest: Allen and Unwin, 2006), 163–164.

41 Ibid., 197.

'fog-filled' streets.[42] In this way *Candy* aligns with other male drug writing in its portrayal of a controlling and apocalyptic sense of place. This is most evident in the urban industrial portrayals of space in Davies' novel. At the height of their drug addiction, the male protagonist and Candy live in a warehouse:

> … twelve hundred square feet of wooden floors and drafty open spaces, a cold place at the best of times. There were only two ways of keeping warm. One was heroin, the supreme crucible in which all sensation melted to a base level of comfort. This was generally how we did it. The second way, when there was no heroin, or not enough, was a two-bar heater that we huddled around …. The rest of the warehouse was empty space.[43]

In contrast to the vastness of the 'twelve hundred square feet' of space, the subject is minimalised, deemed insignificant. Heroin, analogised as a 'two bar heater,' is contrasted to the overwhelming size of the space, a seemingly useless defence against the void. This depiction of the heroin trope magnifies the sense of hopelessness of the alienated individual. Like the external space of the warehouse they inhabit, Davies's depiction of the drug using subjects is one that is emptied or emptying. In a natural eco-sense, the warehouse is a symbolic representation of dystopia where the natural progression of time, night and day, is no longer discernible. While there is a subversive, radical use of the drug trope that reveals or enunciates a (male) alienated subject, it is one that in a Kantian sense separates the subject from space and time.

Like Burroughs, Davies' use of the drug trope to destabilise concepts of space is disastrous for the subject-body that occupies that space. Alternative spaces in *Candy* become a political bulwark stripping the subject of agency. Heroin use in *Candy* perpetuates the imprisoning analogies of space and the subject that resonate in other addiction discourse. As with other discursive constructions of drug use, Davies' junk metaphor is one where the subject 'is out of control' – out of control of the societal constraints and constructs, but also out of his own control of self, left disempowered. Davies' use of the drug trope is nihilistic, entrapping rather than enabling. The male subject, normally privileged in the symbolic order, is removed to the periphery in drug literature. It is a shift of subversion but one that also comes with anxiety, depicting agency in crisis. Removed from his symbolic privilege, the male drug addict becomes invisible to the culturally constructed notions of space and time. The protagonist of Davies' novel diarises the loss of self in conjunction with the

42 Ibid., 168.

43 Ibid., 156.

onslaught of addiction and meaninglessness, again using natural imagery to contrast the urban confines:

> Much later: Melbourne, winter.
>
> My day in the light, the day is darkening. I'm hurling all the little joys against the greater sadness. The sadness is a giant weight. It presses down. Its meaning: 'What's the point?' The little joys are pebbles. The pebbles are getting smaller and smaller and the weight of the sadness is growing, the sadness is gaining density and mass, until in the end I'm throwing handfuls of dust at matter so thick there's no space between the molecules. Nowhere anywhere for anything to move. The years roll on.[44]

While time takes on material spatial density, the junk collapses time in Davies' novel, reducing the protagonist's own ontological need, desire, and even dilemma. While there is a sense of subversion in what Baykan discusses in the following chapter as a collapsing of spatio-temporal dualities, it does not depict agency for the subject-body in *Candy*. For Davies' male protagonist, time, like space, is externalised, condensed and overwhelming. It is a depiction of time that exemplifies the conventional perspectives of time that Grosz describes as a 'pervasive force to such an extent that we cannot bear to think it, we prefer that it evaporates into what we can comprehend or more directly control'.[45] Junk time as it is represented in Davies' novel cannot be controlled. Ultimately, because of the pervasiveness of time on the body, the drugged male subject in *Candy* 'evaporates' also.

The perversion of the male addict in Davies' novel is that he ultimately loses himself to spatio-temporal constructs. The junk trope as it is used in *Candy*, destabilises subjectivity to the point of disappearance. The 'addict'-protagonist of Davies' novel states 'I feel I am nothing but a dividing line. I don't know who I am.'[46] In *Candy*, the male junky is 'replaced by space' similar to Grosz's description of Roger Callois' psychotic who 'is unable to locate himself where he should be; He is captivated and replaced by space, blurred with the positions of others'.[47] Davies' use of the junk metaphor inculcates a nihilistic vision of the subject consumed by the very spatio-temporality that he once controlled. In *Candy* the never-ending spatial and temporal expansion of junk equates to

44 Ibid., 13.

45 Grosz, *The Nick of Time: Politics, Evolution and the Untimely*, 5.

46 Davies, 2006, Candy, 327.

47 Grosz, *Space, Time and Perversion*, 30.

an emptying of the male subject – so much so that he becomes invisible. The following excerpt from *Candy* exemplifies the temporal and spatial disappearing body of junk:

> So much of my own life seemed to be spent waiting on park benches, knotted with stomach cramps and trying to look casual, willing the dealer to arrive. Roger was no more than a ghost to me. But it was during my own long waits in parks – they say that if you've been a junkie for ten years, you will have spent seven of them waiting – that I began to realise that I, too, was becoming transparent. That I myself may well have been a ghost, and that I was haunting my own body ....[48]

Davies' representation of the disappearing, ghost-like junky – the panicked individual – is akin to Deleuze and Guattari's (1987) assertions of the 'empty body' of the addict in their account of a Body without Organs (BwO). Grosz suggests that the empty body of the male addict in male drug literature like Burroughs' *Naked Lunch* can be understood in terms of Deleuze and Guattari's (1987) assertions of the Body without Organs (BwO):

> ... subject and object can no longer be understood as discrete entities or binary opposites. Things, material or psychical, can no longer be seen in terms of rigid boundaries, clear demarcations; nor, on an opposite track, can they be seen as inherently united, singular or holistic...The BwO is the body in abundance of its (biological, psychical, and signifying) organization and organs ... 'not a dead body but a living body all the more alive and teeming once it has blown apart the organism and its organization ... the full body without organs is a body populated by multiplicities.[49]

The BwO is an ideal body freed from societal constraint – a body that, through its multiplicity, has potential in 'what it does, how it functions, what it affects, what it produces'.[50] According to Delueze and Guattari, in opposition to the potentiality of the full BwO 'in and through which intensities flow and circulate, where production are engendered,' is the emptied BwO of the drug addict.[51] Delueze and Guattari specifically refer to Burroughs' portrayal of the

48 Davies, *Candy*, 120–121.

49 Elizabeth Grosz, *Volatile Bodies: Toward a Corporeal Feminism* (St Leonards: Allen & Unwin, 1994), 167–168.

50 Ibid., 170.

51 Ibid.

addicted body in *Naked Lunch* to exemplify the empty BwO, a body 'evacuated not only of organs and organization but also of its intensities and forces'.[52] Deleuze and Guattari suggest that the addicted male body portrayed in *Naked Lunch* 'fills itself to the point where nothing further can circulate…the junkie is filled with "refrigerator-waves," with the numbing-enlivening Cold':[53]

> [A junkie] wants The Cold like he wants his Junk – Not OUTSIDE where it does him no good but INSIDE so he can sit around with a spine like a frozen hydraulic jack … his metabolism approaching Absolute Zero.[54]

Burroughs,' like Davies,' ghost-body 'seem to have emptied itself too fast, too definitively…disarrays itself too much, so that it closes in on itself, unable to transmit its intensities differently, stuck in repetition'.[55] An understanding of the (male) addicted body as antithetical to the BwO, can help to explain the stagnation of agency in male drug literature. However, within Delueze and Guattari's assertions of the drug using-subject body, there is a danger of perpetuating the derogatory stereotypes of the addict (and other non-conforming modes of subjectivity) present within dominant discourse. Furthermore, the lack of gender differentiation in their discussion of the BwO, highlights the phallocentric privileging of the male body as desideratum in Delueze and Guattari's work. The representation of the female addict and the possibility of agency in female drug literature problematises Delueze and Guattari's notion of the emptied BwO of the addicted body.

In order to cope with the ontological loss, the male protagonist in *Candy* invents a world via the trope of drug use that can be viewed, in an Irigarayian sense, as a way of sustenance and survival: 'In order to recapture that whole sensation of the inside of a body, he will invent a world'.[56] This world, however, is one that erases the female character in order to sustain the hollowed bodies of the male junky. Amidst the loss of self and the mere survival in a threatening and overwhelming time and space, the female protagonist is not a parallel sufferer. Candy, the protagonist, is a possession – she is objectified by the male protagonist/narrator (along with place and time) and is separated/cut off

52 Ibid.

53 Ibid., 170–171.

54 Gilles Deleuze and Felix Guattari, *A Thousand Plateaus* (Minneapolis: University of Minnesota Press, 1987), 153–154.

55 Grosz, *Volatile Bodies: Toward a Corporeal Feminism*, 171.

56 Luce Irigaray, *Elemental Passions*, trans. Joanne Collie and Judith Still (New York: Routledge, 1992), 12–15.

via the drug trope. Candy is synonymous to the drug – she is both heroin and heroine. She sustains the drugged body of the male protagonist and maintains the world he creates to survive his ontological loss. There is no agency for the female subject in *Candy* – just as there is no autonomy of the temporal or spatial surrounds from the ontological authority of the male protagonist/narrator. The male protagonist not only voices her story, in Groszian terms, he builds his 'universe' upon the 'erasure' of her 'body and contributions' as support for his 'hollowed space.'[57] This is evidenced in the representation of Davies' female protagonist as prostitute, as mother, as wife. The following excerpt exemplifies Candy's mothering/nursing of the male protagonist when the heroin they buy is 'cut' and ineffective:

> Candy's bathing my face with warm water and a tea-towel. Even though she is sick, she is loving and gentle, and I love her. She says things like, my poor baby my poor baby it'll be all right. I keep wincing. Even the warm water hurts.[58]

Such representations show the designation of women as what describes as:

> the guardians of the private, the interpersonal' a disavowal that allows men to appropriate 'conceptual and material worlds... the right to a place or space... correlative with men's seizure of the right to define and utilize a spatiality that reflects their own self-representations.[59]

Just as space and time are externalised and objectified in Davies' novel, so too is the female subject.

While Davies' use of the drug trope re-imagines spatial and temporal exigencies upon the subject-body, it also magnifies an ontological impasse and futility of the male character that situates himself as separate from space and time in his attempts to have control. In Davies construction, the drug trope immobilises and pacifies the male protagonist in *Candy* who is emptied by his ontological dilemma. Antithetical to political subversion or alternatives, the drug trope as it is used in Davies' novel disenfranchises the subject and strips agency in a dystopian dislocation of the subject from space and time.

57 Elizabeth Grosz, 'Women, Chora, Dwelling,' *Postmodern Cities and Spaces*, eds. Sophie Watson and Katherine Gibson (Oxford: Blackwell, 1995b), 55.

58 Davies, *Candy*, 17.

59 Grosz, 'Women, Chora, Dwelling,' 55.

5 'Sky a Thousand Miles High':[60] Transcendence, Opportunity and Celebration in Garner's *Monkey Grip*

In contrast to the disconnect portrayed in Davies' novel, *Monkey Grip* evokes a sense of the subject in space that theoretically aligns more closely with Baykan's use of Delueze and Guattari's rhizomatic space of in-betweens 'in terms of an open-ended, dynamic space that connects elements in constant flux'.[61] The use of the drug trope in Garner's novel re-situates the spatio-corporeality of the female subject-body as fluid. Furthermore, in line with Collins analysis of the respatialisation of the subject-body via keening, the drug trope is used in Garner's novel to evoke a sense of celebration from the peripheries. Respatialisation as it is examined by Baykan and Collins parallels Garner's reconceptualisation of the drugged space as a vantage point that does not see women as marginalised, but rather in a position of power. Peripheral female subjects become politically motivated subject-bodies empowered as they are already on the edge. Julia Kristeva suggests that it is this positionality that makes women dangerous 'dissident[s] … here to shake up, to disturb, to deflate masculine values, and not to espouse them'.[62] Further, Barbara Baynton alludes to the power of the periphery in her discussion of Australian women's writing:

> In its subversive textuality, it functions to deconstruct the 'place' of women in the (male) imaginary. The writer mimes the role imposed upon women by pointing to it from the stance of a dissident, speaking to a tradition from its dangerous margins.[63]

Monkey Grip comes from the dangerous peripheries – a site of feminist political engagement. Garner's use of the drug trope is suggestive of the way in which the subject belongs to space, is 'emplaced' in temporal and spatial surrounds – making space, place. This is evidenced in the way Garner uses the drug trope to re-envision time and space in *Monkey Grip*.

Like Davies, Garner harnesses the drug realist mode to explore temporal and spatial alternatives for the subject. However, while Davies' use of the drug

60 Helen Garner, *Monkey Grip* (Ringwood: Penguin Books, 1978), 171.
61 Baykan, in this volume.
62 Julia Kristeva, ed., *Polylogue* (Paris: Du Seuil, 1977), 148.
63 Kay Schaffer, *Women and the Bush: Forces of Desire in the Australian Cultural Tradition* (Cambridge: Cambridge University Press, 1988), 168.

trope distorts conventional spatio-temporal conventions, the drugged protagonist is left disempowered in a disconnected dystopia. A clear way to demarcate the fictionalised space and time of Garner's and Davies' novels is in the provision of the utopian impulse in their novels. Gelder and Salzman view the fiction like *Monkey Grip* as a portrayal of utopian images – a 'fruitful interaction between politics and fiction, realism and narrative experimentation, characteristic of that earlier period of social ferment'.[64] While *Candy* fragments social constructs of time and place via the drug trope, it does not have the political mobility or hope for social transformation of *Monkey Grip*. Margaret Henderson and Shane Rowlands make a similar distinction.[65] They attribute to Garner's *Monkey Grip* 'a deep sense of political engagement,' something they deem missing from the grunge of the 1990s. They argue that texts like *Monkey Grip* 'represent moments of radical history, when alternative lifestyles were ideological positions rather than consumerist (or fashion) statements'.[66] Unlike the nihilist capitalism evoked in the consumption of heroin as it is portrayed in *Candy*, the drug is used in Garner's novel in a more organically engaged and politically positive way.

Significantly, while Garner's female protagonist, Nora, does use drugs frequently – such as amphetamines, alcohol and marijuana – she does not use heroin herself. In this way there is a more implicit use of the heroin trope in Garner's novel – a reading of how the junk space impacts/inscribes itself upon subjectivity. Like Muramoto's examination in this volume of the spatio-corporeality potential of engawa, and Baykan's examination of the disrupting dualisms of Orlan's performativity, the drug trope as it is used in Garner's novel is antithetical to the Kantian notion of the self as separate from space/nature in Davies' novel. A Spinozan interconnection of self, 'within' nature, is reflected in the omniscience of the junk trope in Nora's narration: there is 'junk oozing in the atmosphere' of Garner's novel.[67]

Antithetical to the opiated drug addict of Davies' novel, Nora uses different drugs for different purposes signifying a sense of control and choice. The heterogeneity of her drug use also parallels the multiplicity and potential of the postmodern subject – a subject that is always becoming – rather than stagnating. Unlike the alienated male protagonist of Davies' novel, Nora does

64 Ken Gelder and Paul Salzman, *After the Celebration: Australian Fiction 1989–2007* (Melbourne: Melbourne University Press, 2009), 262.

65 Kirsty Leishman, 'Australian Grunge Literature and the Conflict between Literary Generations,' *Journal of Australian Studies* 5 (2009): 97.

66 Ibid.

67 Garner, *Monkey Grip*, 79.

not have a fear of isolation – in fact she usually took drugs by herself: 'I was tripping again by myself'.[68] Nora's solitary use of the drugs evokes her sense of autonomy and empowerment. Nora creates a space, a drugged space, where she could fill herself – balance herself – rather than emptying herself. Garner's protagonist describes herself as a 'reservoir' not the hollowed out empty body that is simulated via the drug trope in Davies' novel.[69]

Garner's portrayal of Nora epitomises Grosz's temporal respatialisation of the postmodern subject. *Monkey Grip* is a literary reflection of Grosz's outlook on time as having 'a life of its own, time deviates, splits, divides itself…it is we who are in time, rather than time that is in us; it is time which inhabits us, subsists or inheres within and beyond us'.[70] The drug trope as it is used in *Monkey Grip* destabilises the body/time demarcation to allow time to permeate subjectivity in a non-conventional way. This is evident in the Nora's movement along the current of time directed via the drug trope, which distorts conventional constructs of time:

> I spent another coke night, almost till dawn, in my bed with Bill. We talked about things I had never talked about before: what it means to be alive in 1975, what change is and might be, how we see ourselves fitting in (or not) to this society, what the next step is or might be. We talked about these desolate things…the dawn came and we got up and attended to the children.[71]

In an empowering way the division between night and day are transcended by the coke-using Nora. Also collapsed are the stereotypical depictions of the drug using mother as inattentive to her children's needs. Nora's drug use reconceptualises time – wresting it from conventional allocations of the subject within time. Garner's use of the drug metaphor reframes time and also relocates the subject. The subject-body in *Monkey Grip* can be seen as a literary reflection of Grosz's understanding of the 'ontology of time' – that bodies are continuously in a state of becoming as 'temporal beings.'[72] Nora's state of 'becoming' is represented via natural symbolism:

68 Ibid., 208.

69 Ibid., 146.

70 Elizabeth Grosz, *Time Travels: Feminism, Nature, Power* (Crows Nest: Allen & Unwin, 2005), 3.

71 Garner, *Monkey Grip*, 147–148.

72 Grosz, *The Nick of Time: Politics, Evolution and the Untimely*, 4.

> I was balancing myself out nicely, there, in the quietness...it had no edge or sharpness. It was merely a series of blurred, dull facts. Whenever I woke from dozing, I found the sky as grey, the surf noise as steady, the house as peaceful as they had been when I lay down hours before. I had no idea what the others were doing, nor what time it was.[73]

Instead of the externalised depiction of natural disasters used to evoke the alienated drugged protagonist in Davies' novel, Nora is enmeshed in the pace of natural time in a peaceful way. The drug metaphor portrayed in Garner's novel disrupts conventional constructions of time – instead of trying to harness time, the novel portrays it in a Groszian sense as 'life succumb[ing] to its rhythms, direction, and forces, to the ever pressing forces of development, growth, and decay'.[74] Drugs are symbolic of a temporal transformative experience in *Monkey Grip* portraying the body/subject as part of a system of forces – uncontainable, ever changing and full of agency.

Agency is also evident in the many dream analogies used and examined by Nora to manoeuvre her way through the 'real' temporal and spatial shifts in her life. Like Collins' and Baykan's assertions on liminality in this volume on – the power of the 'in-between' space – the drug trope in *Monkey Grip* is used to portray the temporal and spatial interstices between the dreamed drug spaces of the psyche and the 'real' fragments of urban space where possibilities and potential are revealed. As with Benjamin's 'narcotic dream city,'[75] Nora's dreamscapes provide a site where alternatives can be explored. Nora acknowledges the potential elicited from the dream state in the following scene where she is woken by another drug using friend:

> "Have you used all that coke?"
>
> "I haven't used any of it," I replied, sitting up and putting my hands to her cheeks, a familiarity I probably would not have dared if I hadn't just leapt out of a dream. "Why – do you want some?" We laughed.[76]

Like the drugs themselves, dreams offer a space for other possibilities to be explored. In the following excerpt the real and dream spaces are collapsed into fantasies symbolic of the drug experience:

---

73 Garner, *Monkey Grip*, 242.

74 Grosz, *The Nick of Time: Politics, Evolution and the Untimely*, 5.

75 Sophie Watson and Katherine Gibson eds., *Postmodern Cities and Spaces* (Oxford: Blackwell, 1995), 74.

76 Garner, *Monkey Grip*, 180–181.

> The fantasies were out of control. They were indescribably delightful. The most trivial word or image would trigger off a bout of them. And every time, when I could no longer ignore their total lack of connection with, or meaning in, the real world, the bad moments followed, as reliably as the thump at the end of a coke flash.[77]

Via the drug trope Nora's dreams are interwoven with her spatial connections in a transformative way:

> I dreamed again and again of houses, always big airy open houses, always beside the sea, wind flowing freely through the rooms, people pleasantly disposed or working quietly, and miles and miles of ocean out every window. In such a house of dreams I spoke with Gerald, this unknown figure: though our heads were close together, his voice was indistinct, hard to hear like a bad telephone connection. I woke up and my room was full of sunny wind, red curtain flying and the noise of the market battering away out there past the Ah Change Trading Company.[78]

The sinuous spaces of Nora's dream are mirrored with her 'real' space. Each space equally interconnects with and reflects the transcendence of the drug trope upon the representation of the subject-body in Garner's novel. In opposition to Davies' 'twelve hundred square feet [of] … drafty open … warehouse' that empties the protagonist in *Candy*, Garner's 'big airy open' dream/house/space 'en-habits' and fulfils the subject-body.

The possibilities created in Garner's drugged dreamscapes are also evidenced in her novel's re-conception of city space. This is evident in Garner's transformation of what Gelder and Salzman define as 'a corrupting Melbourne.'[79] Certainly both cities represented in *Candy* perpetuate the urban space as imprisoning habitats – the corrupting Melbourne and the 'decadently tempting' Sydney.[80] The cityscapes in *Candy* are amorphised as the dealer, and Candy the woman is the drug. Instead *Monkey Grip*'s

77 Ibid., 136.

78 Ibid., 137–138.

79 Ken Gelder and Paul Salzman, *The New Diversity: Australian Fiction 1970–88* (Melbourne: McPhee Gribble Publishers, 1989), 103.

80 Ibid.

> inner-urban Melbourne community of musicians, actors, teachers and drop-outs is presented in loving detail ... celebrat[ing] everything from the Fitzroy baths to the theatrical activity of Carlton.[81]

Garner's use of the drug trope to collapse the dream and city spaces re-imagines the potential of the carnivalesque of Italo Calvino's dream-city analogy:

> With cities, it is as with dreams: everything imaginable can be dreamed .... Cities, like dreams, are made of desires and fears ... their rules are absurd, their perspectives deceitful, and everything conceals something else.[82]

As with Baykan's, Collin's and Muramoto's exposition of respatialisation beyond binarised constraint in this volume, Garner's re-visioning of the city space is most evident in the way the novel goes beyond the dichotomies of city/centres and country/frontiers. Urban space becomes rural in Garner's novel where leitmotifs such as fruit trees, gardens and other pastoral depictions of the cityscape collapse borderlines. Garner reframes the urban space via a pastoral ideal, a Spinozan belonging to space. Like other female drug writing such as Anna Kavan's *Julia and the Bazooka and Other Stories* (1975), and to some extent Elizabeth Barrett Browning's poem 'A True Dream' (1833/1914), Garner recreates a pastoralised urban space – an urban dreamscape suggestive of arcadia. This is exemplified in the following movement through inner-city Melbourne:

> "Wanna take the kids to the baths?" he said ... we rounded up [the kids] ... and disentangled our bikes from the heap outside the kitchen door .... We bumped over the gutter and on to the softening bitumen. The kids begin to sing. We roll in unison ... down the wide road and into the green tunnel, the cave of the Edinburgh Gardens. No-one around, though it is ten o'clock in the morning. The hoses flick silver strings on to the drying grass. The cicadas beat a rhythm that comes in waves, like fainting or your own heartbeat. We sweep round the corner into the Belgium Lane, where the air is peppery with the scent of cut timber and even on this still day

81 Ibid.

82 Drusilla Modjeska, ed., *Inner Cities: Australian Women's Memory of Place* (Ringwood: Penguin Books, 1989), 4.

> the poplars flutter over the ancient grey picket fence; they thrust up their sprouts through the cracking asphalt under our wheels .... I lower myself gingerly on to the blazing ground and watch the kids approach the pool. "No-one will ever understand," I say to Clive, "but this is paradise".[83]

The urban/rural collapse of space here highlights a fragmentation of the dichotomies of space in *Monkey Grip*. Garner's use of pastoral idealism, as also is evidenced in the opening passage from the novel:

> In the old brown house on the corner, a mile from the middle of the city, we ate bacon for breakfast every morning of our lives .... At night our back yard smelt like the country.[84]

The imagery of the pastoralised urban creates alternatives evoking 'newness and freedom'[85] that Wright suggests. By collapsing these spatial borderlines other restrictions placed on the subjects that live in those spaces are also minimalized:

> There were never enough chairs for us all to sit up at the meal table; one or two of us always sat on the floor or on the kitchen step, plate on knee. It never occurred to us to teach the children to eat with a knife and fork. It was hunger and all sheer function: the noise, and clashing of plates, and people chewing with their mouths open, and talking, and laughing. Oh, I was happy then. It was early summer ... teetering as many were that summer on the dizzy edge of smack ....[86]

The drug trope as it is used in Garner's novel jeopardises the urban/country split – a portrayal of an interconnectedness of space, time and the body that captures the Spinozan It Is – one that observes nature from within nature, as nature.

In contrast to the subject's relationship with nature in Davies' novel, nature is not objectified in *Monkey Grip* but mirrors the subject-body that becomes enmeshed with it. Garner's use of the natural environment, a Spinozan sense

---

83 Garner, *Monkey Grip* (Ringwood: Penguin Books, 1978), 10.

84 Ibid., 1.

85 Turner, *National Fictions: Literature, Film and the Construction of Australian Narrative*, 25.

86 Garner, *Monkey Grip*, 1.

of being part of the space, is evident in the following passage where the natural imagery mirrors the internal space/psyche of the female protagonist.

> ... when I looked in the mirror, I saw a preoccupied face, a worried head, a body out of sync with the mind ...
>
> ... disconnected fragments tumbled about ...
>
> I realised I had a stream of thoughts about him which ran for the most part below conscious level. I noticed jets spurting up from this stream: comparisons with other relationships I know of which had weathered massive changes and shifts of balance ...
>
> The sky was still cloudy, but the sun was fighting through that clear line at horizon level, laying a strip of light, uninterfered with, across the tops of the ti-tree scrub.[87]

Again natural imagery is used to describe the power of the female subject, one that 'weathers' – is ever-changing and one that also erupts. The 'spurting' imagery of the subject resonates with Cixous' assertions on the volcanic female body one that writes, erupts from the body. This is evoked equally powerfully in the following drug-imbued passage:

> In the morning, light and air wake me. I go outside and see a sky a thousand miles high, covered with a fine net of almost invisible cloud. My head begins to turn, it fills with unspoken words, I don't try to seize them but let them run unchecked. They seem to slip into my veins and my limbs and the capillaries of my skin. It is just convalescence, and the summer morning. The universe resounds the joyful cry "I am?"[88]

Garner's imagery in *Monkey Grip* epitomises at times Cixous' hopes for l'écriture feminine, moments of an unbounded writing from the body, a writing that depicts a female psyche that is freed from spatial and temporal constraint. As Gelder and Salzman say of Garner's style:

> [it is] ... an example of her depiction of a feminine consciousness conveyed through a specific female style. That style may have been far from the wilder reaches of écriture feminine, but it was (and remains) the most impressive aspect of Garner's writing: she never writes an ugly sentence,

87 Ibid., 242–243.

88 Ibid., 171.

> and is always attuned to the luminous nuances of the details of everyday life.[89]

Garner's depictions of the natural environment also align her with the Romantic feminist aspirations of Barrett Browning that also depict a Spinozan connection with nature to contest phallocentric portrayals of women and nature. In the following excerpt from *Monkey Grip* Nora situates herself within/as part of nature in an empowering way:

> I went to the beach by myself in the late afternoon. I rolled up my jeans and stood on the wet sand where the waves barely reached. I stared at my brown feet. I just stood there. I only stayed on the beach for about twenty minutes. Then I climbed back up the ramp and walked very slowly back to the house.
>
> The waves hissed all night.[90]

Garner empowers her female protagonist by repositioning her 'outside-in' the natural environment. The natural imagery of the 'waves hissing' has allusions to the serpents of Eden and Nora reconceptualises paradise and her relationship with it. Outside of the symbolic city Nora finds composure to re-establish in order to go home to the city space of alternatives, the fractures, the multiplicity. Garner ends *Monkey Grip* with the paradox of the postmodern condition. Nora is at once an autonomous agent while also interconnected in spatio-temporal becoming:

> In the middle of the night I woke up and went outside to piss on the grass. The moon hung low in the sky above the quiet hedge. I squatted down at the corner of the house and let my piss run down the bare, grey earth in a trickle. I stood up, wiped myself with my hand, and rinsed my hand under the tap. I stood still, staring at the moon and feeling the soft air on my skin. Claire's car sat there behind me, a big silent bulk in the dark. I though again of her and Javo, and instead of that pain came the thought,
>
> "Well…so be it. Let it be what it is."

89 Gelder and Salzman, *After the Celebration: Australian Fiction 1989–2007*, 180.

90 Garner, *Monkey Grip*, 241–242.

> … In the morning the sky was clear, the sunlight lay on the scrubby grass in long, pinkish-gold strips. The absent-minded carolling of magpies dropped out of the pine trees half a mile away.
>
> Time to go home.[91]

The potential of Garner's novel rests in Nora's journey home – in the novel's reframing and re-conceptualising of centres and frontiers. The political outcome of Garner's novel can be seen as a literary representation of Grosz's hopes for a reconceptualization of space, time and the body. As Grosz suggests, for women to

> occupy another space, or to … occupy this space differently' women need to return 'to those places from which they have been dis-or re-placed or expelled … in order to be able to experiment with and produce the possibility of occupying, dwelling or living in new spaces, which in their turn help generate new perspectives, new bodies, new ways of inhabiting.[92]

The transcendence of Garner's novel is one found through the cracks of space and time – a fragmentary transformation from the peripheries.

6 Conclusions and Possibilities

Along with other themes and leitmotifs of the city, nature, love and relationships, Garner's use of the drug trope in *Monkey Grip* is one that provides a platform for political engagement. The drug trope destabilises spatial and temporal confines creating opportunities to examine alternatives for the individual, in particular women, in Australian society in the 1970s. The political nature of Garner's novel contrasts with the examination of the ontological dilemma of the male subject in Davies' *Candy*. As with other male drug texts like Burroughs,' the drug trope is used in *Candy* to examine more closely a crisis of agency – one that pertains to the anxiety of the normally privileged white male who, as addict, finds himself, or locks himself outside of the symbolic order that the drugs fragment. Davies' internalized focus on the alienated individual is differentiated from Garner's portrayal of the female subject that is already displaced. Women do not 'fit' into the Kantian 'I am' – they do not belong to or are allowed to have ownership of the 'I.' Women's radical potential comes

91 Ibid., 244–245.

92 Grosz, *Space, Time and Perversion*, 124.

from being 'us,' belonging 'with' in the Spinozan sense. This is tested by *Monkey Grip* – with degrees of difficulty. The alternative space/time created by the drug trope in Garner's novel is represented as equally difficult and idealistic for the female subject that inhabits that space. Nevertheless the drug trope in Garner's novel provides room for the subject to experiment, to try out new ways of the body to move uninhibited.

While the drug trope in Davies' novel displaces the male protagonist and further marginalises the female protagonist, the drug trope in Garner's novel 'emplaces' the (female) subject through a process of respatialisation. Aligning with Baykan's, Collin's and Muramoto's projects of reconceptualising spatio-corporeal confines, women's drug writing offers a unique perspective of the subject that is already on the peripheries, already used to being confronting, on the edge. The drug trope in female drug literature such as Garner's novel embraces the sense of the female subject that is not alienated, but empowered by her position outside of the order. It is what Barbara Brooks sees as '[t]he desire to live on the edge, to be always confronting, breaking out. In our maps, we live on the edge, but our concerns are central'.[93] The 'skewed' view from the periphery that is evoked by the drug trope in *Monkey Grip* provocatively portrays the power of writing from the margins. 'Dangerous' ideas evoked by the drug metaphor are pushed to the centre, subverting and fragmenting from the edge, but its impact enables female agency within the centre.

Bibliography

Ahearn, Edward J. 'The Sordid Sublime: Burroughs's Naked Lunch.' *Visionary Fictions: Apocalyptic Writing from Blake to the Modern Age*. New Haven: Yale University Press, 1996.

Ahmed, Sara. *Queer Phenomenology: Orientations, Objects, Others*. Durham: Duke University Press, 2006a.

Ahmed, Sara. 'Orientations: Toward a Queer Phenomenology.' *GLQ: A Journal of Lesbian and Gay Studies* 12.4 (2006): 543–574.

Armellino, Pablo. *Ob-Scene Spaces in Australian Narrative: An Account of the Socio-Topographic Construction of Space in Australian Literature*. Stuttgart: Ibidem-Verlag, 2009.

Bakhtin, Mikhail. *Rabelais and His World*. Bloomington: Indiana University Press, 1984.

93 Barbara Brooks, 'Maps,' *Inner Cities: Australian Women's Memory of Place*, ed. Drusilla Modjeska (Ringwood: Penguin Books, 1989), 39.

Bakhtin, Mikhail. 'Discourse in the Novel.' *Literary Theory: An Anthology*, edited by Julie Rivkin and Michael Ryan. Oxford: Blackwell Publishers, 1998.

Barrett Browning, Elizabeth. *A True Dream*. New York: MacMillan, New York, 1914.

Brooks, Barbara. 'Maps.' *Inner Cities: Australian Women's Memory of Place*, edited by Drusilla Modjeska. Ringwood: Penguin Books, 1989.

Burroughs, William S. *Naked Lunch: The Restored Text*. London: Harper Perennial, 2005.

Cixous, Helene. 'The Laugh of the Medusa.' *Feminisms: An Anthology of Literary Theory and Criticism*, edited by Robyn R. Warhol and Diane Price Herndl. Hampshire: Macmillan Press Ltd, 1997.

Cixous, Helene. 'Sorties.' *Literary Theory: An Anthology*, edited by Julie Rivkin and Michael Ryan. Oxford: Blackwell Publishers, 1998.

Cixous, Helene. 'The Newly Born Woman.' *Literary Theory: An Anthology*, edited by Julie Rivkin and Michael Ryan. Oxford: Blackwell Publishers, 2004.

Davies, Luke. *Candy*. Crows Nest: Allen and Unwin, 2006.

de Spinoza, Baruch. *The Theologico-Political Treatise (1677)*. Translated by R.H.M. Elwes. Charleston: Forgotten Books, 2008.

Deleuze, Gilles and Felix Guattari. *A Thousand Plateaus*. Minneapolis: University of Minnesota Press, 1987.

Derrida, Jacques. *Writing and Difference*. London: Routledge, 2001.

Derrida, Jacques. 'The Rhetoric of Drugs.' Translated by Michael Israel. *High Culture: Reflections on Addiction and Modernity*, edited by Anna Alexander and Mark S. Roberts. Albany: State University of New York Press, 2003.

Foucault, Michel. *Discipline and Punish: The Birth of the Prison*. Penguin, London: 1991.

Garner, Helen. *Monkey Grip*. Ringwood: Penguin Books, 1978.

Gelder, Ken and Paul Salzman. *The New Diversity: Australian Fiction 1970–88*. Melbourne: McPhee Gribble Publishers, 1989.

Gelder, Ken and Paul Salzman. *After the Celebration: Australian Fiction 1989–2007*. Melbourne: Melbourne University Press, 2009.

Gilbert, Sandra M. and Susan Gubar. *The Madwoman in the Attic: The Woman Writer and the Nineteenth-Century Literary Imagination*. 2nd ed. New Haven: Yale University Press, 2000.

Grosz, Elizabeth. *Sexual Subversions: Three French Feminists*. St Leonards: Allen & Unwin, 1989.

Grosz, Elizabeth. *Volatile Bodies: Toward a Corporeal Feminism*. St Leonards: Allen & Unwin, 1994.

Grosz, Elizabeth. *Space, Time and Perversion: Essays on the Politics of Bodies*. New York: Routledge, 1995a.

Grosz, Elizabeth. 'Women, Chora, Dwelling.' *Postmodern Cities and Spaces*, edited by Sophie Watson and Katherine Gibson. Oxford: Blackwell, 1995b.

Grosz, Elizabeth. 'Women, Chora, Dwelling.' *Gender Space Architecture: An Interdisciplinary Introduction*, edited by Jane Rendell, Barbara Penner and Iain Borden. London: Routledge, 2003.

Grosz, Elizabeth. *The Nick of Time: Politics, Evolution and the Untimely*. Crows Nest: Allen & Unwin, 2004.

Grosz, Elizabeth. *Time Travels: Feminism, Nature, Power*. Crows Nest: Allen & Unwin, 2005.

Hanson, Glen R., Peter J. Venturelli and Annette E. Fleckenstein. *Drugs and Society*. Sudbury: Jones and Bartlett Publishers, 2009.

Irigaray, Luce. *Speculum of the Other Woman*. Translated by Gillian C. Gill. Ithaca: Cornell University Press, 1985.

Irigaray, Luce. Elemental Passions. Translated by Joanne Collie and Judith Still. New York: Routledge, 1992.

Kant, Immanuel. 'Critique of Pure Reason.' *Philosophy on the EServer*. Ames: Iowa State University, 2014. Viewed 25 July 2014. http://www.eserver.org.

Kant, Immanuel 'Considering the Ultimate Ground of the Differentiation of Directions in Space.' *Theoretical Philosophy*. Translated by David Walford. Cambridge: Cambridge University Press, 1992.

Keane, Helen. *What's Wrong with Addiction?* Melbourne: Melbourne University Press, 2002.

Kristeva, Julia, ed. *Polylogue*. Paris: Du Seuil, 1977.

Kristeva, Julia. *Powers of Horror: An Essay on Abjection*. Translated by Leon S. Roudiez. New York: Columbia University Press, 1982.

Kristeva, Julia. *The Kristeva Reader*, edited by Toril Moi. New York: Columbia UP, 1986.

Kristeva, Julia. 'Women's Time.' *The Feminist Reader: Essays in Gender and the Politics of Literary Criticism*, edited by Catherine Belsey and Jane Moore. London: Macmillan, 1989.

Lever, Susan. *Real Relations: The Feminist Politics of Form in Australian Fiction*. Sydney: Halstead Press, 2000.

Marlowe, Ann. *How to Stop Time: Heroin from A to Z*. London: Virago Press, 2002.

Merleau-Ponty, Maurice. *The Structure of Behavior*. Translated by Alden L. Fisher. Boston: Beacon Press, 1963.

Merleau-Ponty, Maurice. *Phenomenology of Perception*. Translated by Colin Smith. London: Routledge Kegan and Paul, 2002.

Modjeska, Drusilla, ed. *Inner Cities: Australian Women's Memory of Place*. Ringwood: Penguin Books, 1989.

Palmer, Cynthia and Michael Horowitz, eds. *Sisters of the Extreme: Women Writing on the Drug Experience*. Rochester: Park Street Press, 2000.

Plant, Sadie. *Writing on Drugs*. London: Faber and Faber, 1999.

Prowse, Nycole. '"Tracking" Cixous's Medusa? A Comparative Analysis of Literary Representations of Women and Their Addiction.' *Cultures of Addiction*, edited by Jason Lee. New York: Cambria Press, 2012.

Schaffer, Kay. *Women and the Bush: Forces of Desire in the Australian Cultural Tradition.* Cambridge: Cambridge University Press, 1988.

Shildrick, Margrit. *Leaky Bodies and Boundaries: Feminism, Postmodernism and (Bio) Ethics*. London: Routledge, 1997.

Stallybrass, Peter and Allon White. *The Politics and Poetics of Transgression*. London: Methuen, 1986.

Turner, Graeme. *National Fictions: Literature, Film and the Construction of Australian Narrative*. Sydney: Allen & Unwin, 1986.

Watson, Sophie. 'Fragmented Visions.' *Inner Cities: Australian Women's Memory of Place*, edited by Drusilla Modjeska, Ringwood: Penguin Books, 1989.

Watson, Sophie and Katherine Gibson. *Postmodern Cities and Spaces*. Oxford: Blackwell, 1995.

Welsh, Irvine. *Trainspotting*. London: Vintage, 2010 [1993].

CHAPTER 2

Respatialising the Body: The Ontologically In-between Subject in Orlan's Body of Work

Burcu Baykan

Abstract

This chapter explores the respatialisation of the embodied experience of space through the French multimedia and performance artist Orlan's body and identity altering practices. By primarily focusing on the artist's multifaceted surgery-performance series, The Reincarnation of Saint Orlan (1990–1993) and her subsequent series of digital self-portraits, Self-Hybridizations (1998–2007), this chapter traces the complex relationships between the human and the non-human domains that appear in her work, as well as the mutating, in-between bodily space that is configured within these meetings and crossovers. Thus, the main intent of this chapter is to engage with and explore these dynamic, unstable and transient states of being that Orlan's work reflects; the metamorphic, in-between areas related to the understanding of self in ontological, philosophical and artistic sense. The investigation undertaken here for this purpose primarily draws on Deleuze and Guattari's formulation of the body through their theory of becoming-other.

1 Introduction

This chapter creates a shift in terrain by turning from literature to performance art. As such, it delves into the respatialisation of the embodied self through the French performance and multimedia artist Orlan's body and identity altering practices. Drawing upon Gilles Deleuze and Félix Guattari's concept of *becoming-other*[1] and its inherent set of vocabularies as a critical lens, this study explores a series of corporeal respatialisations, through which

1 Gilles Deleuze and Félix Guattari, *A Thousand Plateaus: Capitalism and Schizophrenia*, trans. Brian Massumi (Minneapolis and London: University of Minnesota Press, 1987). Gilles Deleuze (1925–1995) is a prolific French cultural theorist in the canon of continental philosophy who has also co-authored with French psychoanalyst and political activist Félix Guattari (1930–1992).

 | DOI 10.1163/9789004365520_004

the human body continually transforms beyond itself in Orlan's interdisciplinary visual art practice. This results in an investigation of two of the major projects by the artist, *The Reincarnation of Saint Orlan* (1990–1993) and *Self-Hybridizations* (1998–2007). Mutative reconfigurations of the body – whether on actual physical level or through digital image technologies – are fundamental to these artistic projects. It is in this respect that Orlan's explorations into the reconceptualisation of the embodied experience of space allude to the reconsideration of corporeal possibilities and potentials by the drug trope in female literature, as it is discussed by Nycole Prowse in the opening chapter of this volume.

By primarily focusing on the artist's multifaceted surgery-performance series, *The Reincarnation of Saint Orlan* and her subsequent series of digital self-portraits, *Self-Hybridizations*, this chapter traces the complex convergences of human and non-human technological constructs, as well as the mutating, in-between bodily space that is configured within these meetings and crossovers. Examining the corporeal interventions explored by Orlan, the main intent of this chapter is to investigate how organic human and inorganic technics are put into contact with each other and the impact of these entangling processes on the creation of in-between spaces – the dynamic, metamorphic and unstable areas related to the understanding of *self* in an ontological, philosophical and artistic sense. It is in this way that, the analysis pursued here is in resonance with Katsuhiko Muramoto's discussion of Teturo Watsuji's 'ethics of in-betweeness' in this volume,[2] as well as this section's general examination of the 'interconnectedness' of the technologies of body, space and nature. For the investigation undertaken in this chapter, I develop a model of the body primarily through Deleuze and Guattari's notion of *becoming-other*. The chapter also includes their concept of *intermezzo* as related to the theory of becoming-other in order to consider Orlan's corporeal reconceptualisation that engenders proliferating connections and non-dualistic forms of bodily space that are no longer based on binary divisions. As such, this chapter is an attempt to elaborate the significance of Orlan's multi-media practices along the lines of Deleuze-Guattarian theory, as well as to explore the points of overlap and resonances between their projects. The current chapter, therefore, extends some of the notions and arguments explored in Chapter 1, where Prowse discussed some of the significant ways in which drugs might impact on and collapse the spatio-corporeal boundaries in literature, and provide a ground where alternative possibilities for the embodied subject can be opened up, via

2 Katsuhiko Muramoto, 'In Search of a Relational Body: Reflections on Sustainable Architecture,' in this volume.

a Deleuze-Guattarian discussion on the body and their notion of *Body without Organs*. But, before plunging into such an analysis, the chapter first briefly describes the art projects that are the focus of this chapter and verbally visualise the elaborate projects Orlan constructs. It then moves on to discuss the implications of the diverse interpenetrating relationships between various human, non-human and technological parts that appear in her works for the comprehension of bodily transformations and alterations.

2 *The Reincarnation of St. Orlan*

Beyond her early work, *The Reincarnation of St. Orlan* is probably the most recognised and extensive project Orlan has undertaken, in which she performed while a cosmetic surgeon operated on her. In this controversial series, where the operating room became her performance space and her own flesh the artistic material, Orlan employed cosmetic surgery for the first time as an art form[3] and underwent a series of nine surgeries to achieve a comprehensive self-transformation. In that sense, she continued to explore the themes present in her ouevre from the onset of her career – the themes of femininity, beauty, renewal and identity change – and use her body to provoke public reaction and debate. Regarding her utilisation of her own corporeality throughout her practice, which takes the form of performances, multi-media installations and photographic pieces, Orlan comments as follows: 'I have always regarded my female body, my artist's body as a particularly suitable material for the production of my work.'[4]

As a multi-media conceptual project consisting of performance art, medicine, biotechnology, prosthetic technologies, communications media, filming and broadcasting, all of the surgery-performances that comprise *The Reincarnation of St. Orlan* were part of public display and carefully designed and choreographed by the artist. Orlan chose to remain conscious under local anaesthesia since her conscious participation was essential to choreographing and performing during the operations,[5] which involved interactive communication with an international audience in art galleries around the globe. Her performance space, which she called *operating-theatre,* had a carnivalesque

3 Orlan, 'Intervention,' *The Ends of Performance*, ed. Peggy Phelan and Jill Lane (New York and London: NYU Press, 1998), 324.

4 Orlan, qtd. in Anja Zimmermann, '"Sorry for Having to Make You Suffer": Body, Spectator, and the Gaze in the Performances of Yves Klein, Gina Pane, and Orlan,' *Discourse* 24.3 (2002): 28.

5 Kate Ince, *Orlan: Millennial Female* (Oxford and New York: Berg Publishers, 2000), 63.

atmosphere. It involved the use of opulent sets and decoration, music, designer costumes worn by Orlan and the surgeons, combined with background dance-performances and multimedia displays, while the bloody procedure was going on.

In relation to the key aims, concerns and motivations that propelled her for embarking on such a project, Orlan frequently states that she is not against cosmetic surgery but opposed to the 'standard criteria of beauty that cosmetic surgery imposes on female and male bodies'.[6] She speaks in detail about her intended rejection of the normative morphologies reproduced by conventional uses of cosmetic surgery. Hence, her performative-surgical project emerges as a critique of the Western beauty standards imposed particularly on women, by means of employing the medium of cosmetic surgery for the first time against its intended purposes – as a means of transformation of oneself.[7] Orlan, in that sense, joins the previous chapter's objects of focus – Helen Garner's and Luke Davies' novels – in her corporeal investigation, insofar as she throws a serious challenge towards the notion of body coded by cultural, in particular, patriarchal discourses. As such, Prowse's argument that the reconceptualisation taking place in both novels might destabilise 'the constraints of spatio-temporal constructs upon the (female) subject',[8] seems to be in line with the way Orlan works to disrupt the culturally accepted standards and expectations of beauty that are perpetuated by patriarchy. Yet, even though the critique of the essentialised notions of beauty in contemporary Western society is one of the most obvious tasks of Orlan's project, her motivations for undertaking these surgeries are much more profound than that. As such, the artist refuses to limit her investigation to purely these paradigms and considers the feminist aspect as 'just a fraction of her work',[9] which in fact, explores more complex issues regarding the human body and identity. Defining her performance surgeries as 'Carnal Art,' Orlan says her 'work and its ideas incarnated in flesh questions about the status of the body in our society and its evolution in future generations via new technologies and upcoming genetic manipulations'.[10]

6 Orlan, 'Interview,' *The South Bank Show.* dir. David Wiles. ed. and pres. Melvyn Bragg. London Weekend Television, 1997.

7 In relation to this, Orlan states: 'I am the first artist to use surgery as a medium and to alter the purpose of cosmetic surgery.' Orlan, 'Intervention,' 324.

8 Nycole Prowse, 'Possibilities from the Peripheries: Time, Space and the Body in the Australian Drug Writing of Helen Garner and Luke Davies,' in this volume.

9 Orlan qtd. in Robert Ayers, 'Serene and Happy and Distant: An Interview with Orlan,' *Body Modification*, ed. Mike Featherstone (London: Sage, 2000), 180.

10 Orlan, 'Intervention,' 319.

She defines another aim of hers as questioning the taboos pertaining to the violation of the bodily integrity in Western culture, by not feeling 'threatened by this collective, ancestral fear of interfering with the integrity of the body ... that the body, formerly viewed as God's masterpiece, is sacred and untouchable, not to be transformed.'[11] In addition to Orlan, writer Jane Goodall neatly expresses the motivation behind her surgical piece: 'Orlan's determination to interfere with her given identity provides the sustaining motivation for her whole *oeuvre*.'[12] In a similar examination, critic August Ruhs characterises her project as follows: 'Orlan irreversibly alters her physical and social identity – she literally sacrifices it to art through plastic surgery, so that she transcends the limits of traditional body art in an unprecedented way – hardly any other artist achieves this aim as radically as she does'.[13] Hence, it follows from all of the above that, at the heart of Orlan's performative Carnal Art project is her investigation of the capacities, limits and mutability of the body and identity through the advances in biomedical sciences and various forms of contemporary technology.[14]

The entire project of *The Reincarnation of St. Orlan,* which investigated such possibilities of bodily manipulation and mutation through new technologies and medical interventions, took place over an extended period of time, spanning from 1990 to 1993. Each of Orlan's operations was designed to alter a specific feature and had a theme, including: *Carnal Art, This is My Body-This is My Software, I Have Given My Body to Art* and *Identity Alterity*.[15] Based on the removal (incision, cutting, extracting and reduction) and addition (implanting, injection, suture), these series of transformations included facial modification, insertion of prosthetic and silicone implants, liposuction from her thighs and reshaping of her ankles, knees, hips, waist and neck.[16] These radically

11 'Harlequin Coat,' *Sk-Interfaces: Exploding Borders – Creating Membranes in Art, Technology and Society*, ed. Jens Hauser (Liverpool: FACT and Liverpool University Press, 2008), 83.

12 Jane Goodall, 'An Order of Pure Decision: Un-Natural Selection in the Work of Stelarc and Orlan,' *Body Modification*, ed. Mike Featherstone (London: Sage, 2000), 155.

13 August Ruhs, 'Orlan Metamorphoser,' *Body as Membrane* nv (1996): 46.

14 The images of Orlan's performance-surgery series might be seen on the artist's official website. Viewed on 5 August 2015, http://www.orlan.eu/works/performance-2/; and http://www.orlan.eu/works/photo-2/nggallery/page/2.

15 Kathy Davis, '"My Body is My Art': Cosmetic Surgery as Feminist Utopia?' *The European Journal of Women's Studies* 4.1 (1997): 26.

16 C. Jill O'Bryan, *Carnal Art: Orlan's Refacing* (Minneapolis: University of Minnesota Press, 2005), 15.

disfiguring and refiguring processes, however, neither ended in an integrated whole or a unified beauty, nor were intended to create such a result. Rather, Orlan acquired some discordant new features, a type of fragmented, uncertain and ambiguous appearance that resists any fixed definition or coherent meaning. Arguing that the artist looks 'distinctly unnatural in appearance,' performance studies scholar Philip Auslander states that 'the work's critical edge derives from the failure of the subject to become the desired image.'[17] With her facial disfigurement and 'failure … to become the desired image,' particularly with the forehead implants she had on her temples to create two horn-like bumps, Orlan deviated from the human body's normative morphology and socio-cultural norms of ideal self. Therefore, the artist became the direct result of her performances; she converted her corporeal space literally into a malleable artwork, into a 'site of public debate' by continuously refashioning her actual physicality.[18] Accordingly, the artist's experimental attitude towards the body, by treating it as mutable material that could be modelled and remodelled infinitely, visualised her embodied space in a constant process of transformation and change throughout the duration of the surgical performances. In that sense, the corporeal respatialisation at stake here has strong resonances with Garner's reconceptualisation of the drugged space in her novel, as examined by Prowse in the preceding chapter. While presenting a different slant on the theme of respatialisation, the spatio-corporeal fluidity and permeability of the female subject-body engendered by the use of drug trope in Garner's *Monkey Grip,* nevertheless, are equally at play in *The Reincarnation of Saint Orlan.* Beyond these physical alterations Orlan experienced in the spatio-temporal coordinates of her body, the images of her mutating flesh were further translated into digital representations and disseminated across the electronic space during the surgeries by means of digital networks of satellite and real time transmission. This, arguably, enabled her body to experience further spatial changes and become even more exteriorised. In that respect, the prominent body art scholar Amelia Jones rightfully observes that the artist is 'enacting herself (and literally *rearranging* her body/self) through technologies of representation as well as medical technology, her body/self is experienced (both by herself and by her audience) in and through technology'.[19]

17 Philip Auslander, *From Acting to Performance: Essays in Modernism and Postmodernism* (London: Routledge, 1997), 131.

18 Orlan, 'Carnal Art Manifesto,' viewed 20 February 2010, http://orlan.net.

19 Amelia Jones, *Body Art/Performing the Subject* (Minneapolis: University of Minnesota Press, 1998), 227; Orlan's official portrait can be seen in the following link: https://sumnonrabidus.files.wordpress.com/2011/02/orlan1.jpg.

3 *Self-Hybridizations*

In her subsequent series of self-portraits entitled *Self-Hybridizations*, Orlan continued to deal with the body and its experimental modifications and mutations, this time using digital technology instead of the scalpel to transform the body. Hence, in this section, I further my discussion on *Self-Hybridizations* series, a complementary body of work to Orlan's surgery-performances. Exploring similar themes such as beauty and identity, this series of computer-generated self-portraits were created by digitally combining the photographs of her surgically altered face with the signifiers of beauty of non-Western cultures: images collected from various pre-Columbian, American-Indian and African sculptures and masks. For *Self-Hybridizations*, the artist worked with a digital imaging specialist and utilised 3-D modelling techniques to produce digital photographic portraits in which her modified facial features were further amalgamated with the stylisations, ritualistic deformations and scarifications of these civilizations.[20] The resulting images reveal the multiplying potential of transformation made possible by new technologies in the digital realm. They offer many variations; Orlan is digitally morphed into what appears to be female figures with bizarre protuberances, enlarged noses, deformed skulls and tattoos.[21] In relation to these photographic pieces, which blend her existing facial coordinates with the primitive disfigurements associated with past tribes, Orlan asserts that they are the 'images of mutant beings whose presence is thinkable in a future civilisation that would not put the same pressures on bodies as we do today'.[22] Indeed, most of the images in the series confound the viewers' expectations of the human corporeality, with their obscured forms such as swollen foreheads, elongated necks, extremely distorted noses and inscribed skin. Combined with Orlan's own modification – the bumps on her forehead – their type is difficult to distinguish.[23]

20 Christine Schultz-Touge, *Orlan: Carnal Art* (Paris: Flammarion, 2004), 166.

21 So far three series have been realised, as Eugenio Viola discusses in detail: 'In Self-hybridations *précolombiennes* (1998–2000), the artist used morphing software to hybridize her new face with masks and votive statuettes of the Olmecs, Mayas, Incas, and Aztecs. The Self-hybridations *africaines* (2000–2003) were obtained by mixing a portrait of the artist with ethnographic photographs of African tribes…In the Selfhybridations *amérindiennes* (2005–2007), Orlan hybridized her own image with portraits belonging to various American tribes.' Orlan, 'The Narrative,' *Orlan: The Narrative*, ed. Lorand Hegyi, Joerg Bader, Eugenio Viola, Marcela Iacub, Donald Kuspit, Peggy Phelan and Orlan (Milano: Edizioni Charta, 2007), 45–47.

22 Orlan qtd. in O'Bryan, *Carnal Art*, 134.

23 The images of Orlan's *Self-Hybridizations* series are available on the artist's official website. Viewed on 5 August 2015. http://www.orlan.eu/works/photo-2/nggallery/page/2.

Since the possibilities of self-transformation are explored through digital manipulation techniques and computer morphing technology, with this particular project, Orlan was able to eliminate the high risks or threats involved in experimenting with the real body tissue. Yet, due to the very same reason, it could also be argued that the pieces that comprise *Self-Hybridizations* series might, in fact, be lacking the depth of her surgical alterations in her *Reincarnation* project, insofar as these works are not based on the transformation of the material flesh, as explained by the artist herself: 'Contrary to what I did with surgical operations, the self-hybridisation series do not inscribe the transformations in my body (my phenomenical body) but in the pixels of my virtual body, mixed with non-organical matter and my own representation, itself reworked by surgery'.[24] Nevertheless, the malleability and permeability of the body achieved throughout this project, are sufficient to create a serious interrogation of human subjectivity, by highlighting, as Orlan says, 'multiple, evolving, mutating identities.'[25]

As this discussion has indicated, the radical disfigurings and refigurings of Orlan's embodiment brought about by cosmetic surgery and more recently by digital morphing technology are unique works of metamorphosis, which seriously complicate the understanding of body's relation to space. Lying on the operating table with her flesh cut open, Orlan makes her body entirely mutable and prompts a radical rethinking of the basic assumptions about the stability and the unity of the corporeal self. In a similar manner, her digital self-portrait series significantly questions the fixed and the static codes that structure the body in conventional humanist configurations. Her Carnal Art projects, therefore – whether digital or material – present a consistent preoccupation with the theme of bodily transformation that evokes an understanding of corporeal respatialisation. As these projects move towards complex spaces where the human body is depicted as a melding of biology and technology, the clear distinctions between the spaces that belong to self and other, human and non-human, cease to exist. Dominant identity categories are disrupted and fixed, organic bodily boundaries are contested. Amidst such boundary-breaking encounters, Orlan's work points out to a need to rearticulate the changing perceptual limits of the bodily space. In confronting these amalgamations of organic human flesh and inorganic prostheses during the performative-surgeries, how is one to articulate the human self? What is it that emerges within the complex interchanges between various human and non-human parts, the flayed flesh

24 Ibid., 134.

25 Orlan, qtd in Peg Zeglin Brand, 'Bound to Beauty: An Interview with Orlan,' *Beauty Matters*, ed. Peg Zeglin Brand (Bloomington: Indiana University Press, 2000), 305.

and technological apparatus? How does this digital blending of various human identities that belong to different civilisations of different times reflect back on the body, on the creation of the individual? Having constructed a detailed analysis of Orlan's complex body projects, the following section critically examines Deleuze and Guattari's formulation of the body through their theory of becoming-other and its attendant vocabularies as constituting the underpinning framework for addressing these various questions that accompany the artist's works.

4 A Deleuze-Guattarian Perspective towards the Body: Subject as Becoming-other

In terms of identity, Deleuze's philosophy can be seen as a critical attempt to leave behind the traditional notion of the subject with any fixed essentiality. In his single authored works, as well as his collaborations with Guattari in their two central texts, *Anti-Oedipus* and *A Thousand Plateaus; Capitalism and Schizophrenia*, he rigorously criticises all self-bounded, autonomous and static forms of *being* that have no connections to their outside. His philosophy is, rather, grounded on a more open-ended, processual and relational ontology at the base of which are the dynamic processes of *becoming*. Accordingly, for Deleuze, 'the struggle for [modern] subjectivity presents itself, therefore, as the right to difference, variation and metamorphosis'.[26] Becomings, in this sense, are processes instigating the transformation of self – not transformation from one static state of being to another – rather they imply a processual metamorphosis, perpetual transition or a constant differentiation that destabilises the fixed, static and unitary understandings of identity. In short, the notion denotes the very dynamic of change, as Deleuze and Guattari write: 'Becoming produces nothing other than itself ... What is real is the becoming itself, the block of becoming, not the supposedly fixed terms through which that which becomes passes'.[27] As thinking oriented towards continual movement and ongoing process, then, the logic of becoming tends towards no spatial end result or a finalised, resettled identity.

Such a process, however, does not materialise all by itself: becomings take place through heterogeneous encounters and multiple connections between different forms and force fields. Rosi Braidotti's work, in dialogue with Deleuze

---

26 Gilles Deleuze, *Foucault*, trans. Sean Hand (Minneapolis: University of Minnesota Press, 1988), 106.

27 Deleuze and Guittari, *A Thousand Plateaus*, 238.

and Guattari's writings, also draws attention to the relational aspect of becoming which is 'about affinities and the capacity both to sustain and generate inter-connectedness',[28] thus, she adds, it is 'external and interrelational.'[29] As a radical form of interconnectivity, it is also crucial to highlight that the theory of becoming involves not only human but also non-human modes of life and their complex interrelationships. It emphasises the shifting and passing interactions between multiple and divergent things, such as human and machine, human and animal, organic and inorganic, and their respective territories. For Elizabeth Grosz, following the work of Deleuze and Guattari, this account cultivates the interconnections between entities 'both human and nonhuman, animate and inanimate, linking organs and biological processes to material objects and social practices',[30] while for Braidotti, it highlights the multiple relational bonds with a variety of 'external forces, of the non-human, inorganic or technological kind'.[31]

Because such a perspective shifts the focus on relationality and connectedness, subjects can no longer be confined to neatly defined ontological systems or individuated *beings* with any fixed spatiality. Instead, they can only be grasped as transitory and connected modes of existence, always *becoming* more and other than themselves through their convergences and alliances with other bodies and external *milieus*. This is to say that subjectivity 'necessarily changes in nature as it expands its connections'.[32] If subjects can pass and slide into each other, interpenetrate with other bodies, then, the strictly polarised, discrete and self-sustaining subject cannot be upheld. This kind of interconnected and interdependent understanding of ontology proves to be central to Deleuze and Guattari's philosophy and is constitutive of their subject, which is in a constant exchange with a series of other biological, social and technical modes of life:

> There is no such thing as either man or nature now, only a process that produces the one within the other and couples the machines together ...

28 Rosi Braidotti, *Metamorphoses: Towards a Materialist Theory of Becoming* (Cambridge: Polity Press, 2002), 8.

29 Ibid., 83.

30 Elizabeth Grosz, *Volatile Bodies: Toward a Corporeal Feminism* (Bloomington: Indiana University Press, 1994), 165.

31 Rosi Bradiotti, 'Affirming The Affirmative: On Nomadic Affectivity,' *Rhizomes* 11/12 (2005/2006): np, viewed on 15 January 2016, http://www.rhizomes.net/issue11/braidotti.html.

32 Deleuze and Guattari, *A Thousand Plateaus*, 8.

> the self and the non-self, outside and inside, no longer have any meaning whatsoever.[33]

From this perspective, bodies and selves are always implicated in their exteriority; through heterogeneous connections and alliances they ceaselessly form with their surrounding terrains. What Deleuze proposes then, in his collaborations with Guattari, is an actively relational, dynamic, shifting and open-ended subject in-process that is 'on the periphery, with no fixed identity, forever decentered, defined by the states through which it passes'.[34] Considering this continual involvement with the external environments, the body as an organically determined, exclusive unity or an ordered, unified whole becomes inoperable in this context. Accordingly, Grosz aptly defines how this account refuses 'to subordinate the body to a unity or a homogeneity of the kind provided by the body's subordination to consciousness or biological organization.'[35] In her discussion of Deleuze and Guattari's notion of becoming, she further remarks that the body is radically reconfigured:

> in terms of what it can do, the things it can perform, the linkages it establishes, the transformations and becomings it undergoes, and the machinic connections it forms with other bodies, what it can link with, how it can proliferate its capacities – a rare, affirmative understanding of the body.[36]

Here, Grosz points to the corporeal model that brings about a dismissal of unifying, homogeneous ground and a focus on the body's capacity to be open to the possibilities of change through multiple linkages and associations.

Finally, this kind of theorisation necessarily entails a certain move away from the binary logics associated with traditional Western thinking, since the body is not defined in its ontological interiority, but rather in its interactions and intertwinements with other bodies and entities. As such, a systematic concern apparent in Deleuze and Guattari's texts is an attempt to undo a series of dualistic structures and categorical orders such as inside/outside, human/machine, human/animal, organic/inorganic and biology/culture among other binary pairs. This results in an expanded sense of ontology, in which all entities

---

33 Deleuze and Guattari, *Anti-Oedipus: Capitalism and Schizophrenia*, trans. R. Hurley, M. Seem and H.R. Lane (Minneapolis: University of Minnesota Press, 1983), 2.

34 Ibid., 20.

35 Grosz, *Volatile Bodies*, 164–165.

36 Ibid., 165.

and things are situated on same field 'which applies equally,' Deleuze and Guattari write, 'to the inanimate and the animate, the artificial and the natural'.[37] This boundary-crossing dimension of the theory of becoming has also been astutely observed by Braidotti, who states that the lines of becoming favour 'the overcoming of the dualism Self/Other" and they 'inevitably and necessarily move into the direction of the "others" of classical dualism.'[38] Having provided the philosophical backdrop of this chapter, the task of the following section is to interweave the theory of becoming outlined here with the practice of Orlan, which demonstrates a strong preoccupation with the production of external relations and connectivities, implying a strong sense of corporeal respatialisation.

5 Theory of Becoming and Orlan's Corporeal Metamorphosis

As a theory that foregrounds the ongoing production of relational encounters between heterogeneous domains or fields, Deleuze and Guattari's becoming-other proves useful for elucidating the complex entanglements of varied human and non-human elements in Orlan's body-based practice, and for drawing out the various degrees of metamorphosis her body undergoes in her Carnal Art projects. In relation to these continual corporeal alterations and the accompanying disruption of the fixed categories of identity in her surgical project, performance scholar Jill O'Bryan argues that the artist's body 'always appears to be in the process of forming' and her identity is 'of one who has exacerbated a state of flux (with surgery) in a live (and therefore already continually changing) human body'.[39] In a similar analysis to O'Bryan, Auslander argues that Orlan's work portrays 'a self for which identity is mutable, suspended, forever in process'.[40] Indeed, the constant disfigurings and refigurings of the artist's body through a series of nine extensive cosmetic surgeries perform the mutability and instability of her bodily identity by marking her transition from a static and fixed state of being to a self-generating and processual one. According to O'Bryan, moreover, they do so 'by reopening the body-boundary which must remain closed in order to guarantee the level of repression necessary for the maintenance of organized subjectivity'.[41] Indeed, Orlan's intimate exploration of her own flesh seriously disrupts the notion of an exclusive, unified human

37 Deleuze and Guittari, *A Thousand Plateaus*, 254.

38 Bradiotti, *Metamorphoses*, 119.

39 O'Bryan, *Carnal Art*, 54.

40 Qtd in O'Bryan, *Carnal Art*, 31.

41 Ibid., 131.

self with any inert nature or stable form of organisation. It discloses the body not as a self-contained, bounded entity with clear delineations between inside and outside, rather as a permeable, transitory and shifting terrain in the public arena.

Accordingly, Orlan's fleeting states of being, which move from one ontological configuration to the next, and her body that always remains in the process of transition throughout the series of surgery-performances become the visual manifestation not of a *being*, but of a *becoming*. As a matter of fact, since the artwork taking place is the very making of Orlan, the artist performs her own becoming – becoming-other than what she is – both in an ontological and artistic sense. To begin this analysis, Deleuze and Guattari's emphasis on becoming as a movement that strives towards no particular end-state or as a pure process that 'produces nothing other than itself'[42] has specific relevance to Orlan's series of surgery-performances that likewise, value process over result. In her statement where she defines her performative surgeries as Carnal Art, Orlan emphasises that:

> the plastic result was not the objective – any result would have served my purpose ... What really matters is the process, what happens in the operating theatre, and the fact that my body has become an area for public debate.[43]

This focus on process and what happens in between – away from a finished and spatialised end product – suggests that Orlan does not aim to achieve or adhere to a reconfigured or finalised form of identity: an effort that would fall back onto a static definition and sedentary being. According to Orlan's own account, her surgical project as well as her digital self-portraits does not imply a reaffirmation of the subject; instead of representing a mere 'constructed re-personalization,' they foreground 'a destabilization of the subject.'[44] Subsequently, this concern with the body as a *temporal process* as opposed to the body as a *spatial outcome* place both of her projects in strong resonance with the notion of becoming that conceives and expresses change in itself.

Deleuze and Guattari's emphasis on the relational aspects of becoming via an open-ended involvement with the outside, and their radical reconfiguration

42 Deleuze and Guittari, *A Thousand Plateaus*, 238.

43 Orlan, 'Carnal Art Manifesto.'

44 Qtd in Rocio De la Villa, 'Orlan's Wardrobe,' *Orlan + Davidelfin: Suture, Hybridisation, Recycling*, ed. Rocio De la Villa and Pedro Alberto Cruz Sanchez (NC: Murcia Cultural, 2009), 32–35.

of the body in terms of contingent connections with various life processes – both human and non-human – provides a powerful means to unpack the complexity of encounters taking place during Orlan's surgical operations. As such, the relational ontology of becoming that Deleuze and Guattari develop, articulates some of the same kind of connections and interactions in which Orlan is interested: interactions that occur between heterogeneous realms – biological and artificial, organic and inorganic, figuration and disfiguration etc. During the surgery-performances, Orlan's skin is literally peeled away from her body and according to the cultural theorist Parveen Adams, in viewing it, we 'find ourselves unhinged in a space that refuses to organize an inside and outside'.[45] The audiences watching the event in the galleries witness the bloody disorganisation and the carnal quality of the flesh, according to Ince, by gaining visual access into the 'fragmented body composed of mass of internal organs, fluids, tissue and muscle' through the incisions and open wounds in Orlan's body.[46] Given this context, Orlan's project fundamentally reverses the relationship of the interior to the exterior by turning the body inside out and, therefore, fractures the spectators' understandings of the body's spatial boundaries. The way Orlan describes her Carnal Art, as 'a way of refiguring yourself, of vacillating between disfiguring and refiguring'[47] further intensifies the destabilisation of neatly organised and strictly bounded corporeal space.

The surgical opening of the artist's body during the performances, moreover, brings her into continuity with other bodies by becoming increasingly permeable to the outside. During the processes of cutting, opening and suturing of her skin, technological parts as various inorganic implants and biomechanical prostheses are inserted and implanted into Orlan's organic flesh, into her bones. Hence, the surfaces of self and other, biological and technological fuse and fold into one another, collapsing the corporeal borderlines. Said another way, as the surgeon's scalpel cuts through the skin, her body is penetrated by other elements, it incorporates them into itself and blends with external parts, provoking an exploration of unfixed corporeal coordinates. Performance scholar Michelle Hirschhorn reflects on the unsettling experience of witnessing these bloody procedures by expressing that there is 'something profoundly destabilizing about watching a woman's face being sliced open, seeing her bleed, seeing medical instruments moving indiscriminately under her skin,

45 Parveen Adams, 'Operation Orlan,' *The Emptiness of the Image: Psychoanalysis and Sexual Differences* (New York: Routledge, 1995), 156.

46 Kate Ince, *Orlan* (Oxford and New York: Berg Publishers, 2000), 52.

47 'Orlan Interviewed' by Hans Ulrich Obrist, in *Orlan: Carnal Art*, ed. Christine Schultz-Touge (Paris: Flammarion, 2004), 199.

that words alone would simply not convey.'[48] Indeed, this is a body with extremely permeable boundaries, a body, according to Grosz, which exceeds its biological designation, 'to open itself up to prosthetic synthesis, to transform or rewrite its environment, to continually augment its power and capacities through the incorporation and into the body's own spaces and modalities.'[49] Accordingly, by subjecting her body to medical technologies, Orlan investigates the possibilities associated with the breakdown of the barriers between internal and external, natural and artificial, human and non-human spaces, as well as their potential couplings. It is in this respect that, my reading of Orlan's live performance comes close to Prowse's reading of the dissolution of spatial boundaries as it is depicted in Garner's *Monkey Grip* in Chapter 1. Common to both pieces, is the creation of spatial connections, interconnections of the self with its various environments, which, in turn, act as a catalyst for the corporeal respatialisations that go beyond binary constructs. While such spatial connections are established through the narcoticising of the urban and the rural, the self and the environment in Prowse's examination, it takes the form of entanglements of the organic human flesh and artificial technological constructs in Orlan's art practice.

To return to the main point of the current argument, within Deleuzian conceptualisation, then, the body becomes an assembled, multiple, dynamic and shifting site of human-machinic interaction. These boundary-crossings and reciprocal exchanges between synthetic and organic elements that open the static homogeneity of Orlan's body to continuous variation, as well as its medical disassembly and reassembly by prostheses, silicone and sutures foreground the body's various technological becomings. The effects of these modifications are in the form of lack, enhancement, displacement or extension, or the assembly of unrelated body parts. These modifications of the body's fleshy territory, in turn, dramatically dislocate Orlan's embodied self from its fixed spatial organisation. On this account, her surgical experiments lead to a radical reconfiguration of her flesh as a heterogeneous site that disturbs traditional spatial boundaries of where the human being ends and the world outside begins. One might say that, searching for an alternative to the view that regards the embodiment as a sealed entity, in Grosz's words, as a 'singular, organised, self-contained, organic body'[50] with clear-cut separations between inside and

48 Michelle Hirschhorn, 'Orlan: Artist in the Post-Human Age of Mechanical Reincarnation: Body as Ready (to be re-) Made,' *Generations & Geographies in the Visual Arts: Feminist Readings*, ed. Griselda Pollock (London: Routledge, 1996), 112.

49 Grosz, *Volatile Bodies*, 187–188.

50 Ibid, 172.

outside, biology and technology, Orlan performs a highly metamorphic self that ongoingly changes by spreading, shifting and overstepping onto other territories, linking to other domains, technologies, spaces in a multiplicity of connections. Amidst such dynamic and complex connectivities, the audiences watching the bloody transfiguration of her flesh in the galleries, simultaneously witnesses the production of an indeterminate, inexact and uncertain corporeal ground. As a body that effectuates its own metamorphosis through the external relations it forms with its various others, Orlan's corporeal self, in turn, is closely aligned with Deleuzian paradigm of the body in its significant move away from a static, unified, self-enclosed space towards a collective site of ongoing interactions. As such, within his formulation, it becomes 'the product of an assemblage – which is always collective, which brings into play within us and outside us populations, multiplicities, territories, becomings, affects, events'.[51]

Making new connections and forging alliances between different fields also appear as the core element of Orlan's self-portrait series *Self-Hybridizations* – not on a material but a digital level. Employing computerised transformation, Orlan, once again, becomes the embodiment of multiple connections between the aesthetics of different spaces and times; intertwinement of ancient and contemporary, visceral and virtual, civilised and primitive, human and non-human dimensions in one physical unit. More precisely, the facial associations she configures between her own surgically altered face and the facial masks of different races and cultures destabilise the separation of individuals into distinct categories, and present their potential fusions that can endlessly be modulated. The convergences and intersections of these different realms of experience, in turn, recreate the body as a complex site of relationality, while opening its personal, homogeneous unity to the possibilities for becoming-other-than oneself. According to Orlan, the images that constitute this project 'have a relationship with the body which is particularly disturbing for us, which completely challenges us and which is very intense... This is the idea of entering into the skin of the other'.[52] By 'entering into the skin of the other', Orlan brings an important dimension into the picture, namely the possibilities of a transitory and interconnected form of existence, the active process of passing and mutating into other identities. In that respect, by means of endless alteration and amalgamation the digital realm offers, Orlan's self, once again, becomes evocative of an uncontained corporeal space that extends

51 Gilles Deleuze and Claire Parnet, *Dialogues*, trans. H. Tomlinson and B. Habberjam (London: The Athlone Press, 1987), 51.

52 Qtd in Ayers, 'Serene and Happy and Distant: An Interview with Orlan,' 177.

towards the territories of other bodies, this time the bodies of other cultures and other time periods. The 'immediate relation with the outside' that Deleuze and Guattari mention as constitutive of ontology,[53] in that sense, seems to find an aesthetic parallel in the artist's digital self-creation. Subsequently, the multiplicity of bodily differences and irregularities at play within these composite images – protruding eyes, skin lesions, enlarged noses and deformed skulls – propels Orlan's embodiment from being mere human to something else, to an open field of becomings that make manifest 'an inhumanity immediately experienced in the body as such'.[54] In the process, like her previous bloody procedures, and the drug trope in Prowse's[55] analysis, these differences steer her bodily image towards an ambiguous, heterogeneous and imprecise state that defies categorical distinctions and ontological assumptions. In order to better address this metamorphic, heterogeneous and indeterminate dimension of corporeality opened up both in *The Reincarnation of St. Orlan* and *Self-Hybridizations,* it is also necessary to invest in the vocabularies of *betweenness* and *intermezzo* that the theory of becoming brings about, notions which are discussed in more detail in the next section.

6 Being *In-between*, Becoming *Intermezzo*

As a model of radical relationality and connectedness, as argued earlier, the theory of becoming also necessitates a certain overcoming of dualistic structures and binary oppositions that have long dominated Western philosophical thought. This emphasis on the non-dualistic continuity between disparate fields and forms introduces a particular aspect of becoming: the strategy of being *in-between.* Deleuze and Guattari's conception of becoming, as a movement that connects and passes between binary terms, 'frees itself from the point, and renders points indiscernible'.[56] The attention, thus, is shifted from the fixed points or states to the movement itself, to the relational passage and transitional states between those terms. Becoming, sweeping away fixed identities, therefore, emerges in a relational *middle*:

> Between things does not designate a localizable relation going from one thing to the other and back again, but a perpendicular direction,

53 Deleuze and Guittari, *A Thousand Plateaus*, 377.

54 Ibid., 273.

55 Prowse, in this volume.

56 Ibid., 294.

> a transversal movement that sweeps one and the other away, a stream without beginning or end that undermines its banks and picks up speed in the middle.[57]

As a logical outcome of these middle spaces, Deleuze and Guattari proceed with their analysis by discussing how becomings take place *in-between* entities or subjects: 'A becoming is neither one nor two, nor the relation of the two; it is the in-between'.[58] Grosz also explains the characteristics of these in-between areas as such: 'the position of the in-between lacks a fundamental identity, lacks a form, a givenness, a nature. Yet it is that which facilitates, allows into being, all identities, all matter, all substance. It is itself a strange becoming'.[59] In sum, as a metamorphic movement between the terms it connects, the space of becoming is 'always in the middle, between things, interbeing, *intermezzo*'.[60]

Because becoming operates in the passing intervals or shifting interstices of ontological orders and categorical distinctions, it provides a way to compose a self that does not differentiate between human and non-human, or biology and technology. That is to say, becomings, which offer ways of inhabiting all sorts of betweens, positions existence in the middle – in the middle of nature/artifice, human/animal, organic/inorganic and so on. This mode of existence in the intermediary spaces, in turn, allows a vision in which 'a dualism or dichotomy can never be assumed'.[61] It, therefore, binds even the most discrete forms, *milieus* and bodies into a dialogue with each other; hence it enables the construction of mobile, non-binary, heterogeneous, blurred arrangements among seemingly divergent elements and distinct forms of life. Within such a context, according to Deleuze and Guattari, the self becomes 'only a threshold, a door, a becoming between two multiplicities'.[62] Accordingly, the following section continues to deploy the theory of becoming as an elucidating strategy to tease out the various transitional, uncertain, in-between corporeal spaces opened up in Orlan's practice – the spaces defined in contradiction to a discretely organised, self-enclosed, organic body.

---

57 Ibid., 25.

58 Ibid., 293.

59 Elizabeth Grosz, *Architecture from the Outside: Essays on Virtual and Real Space* (Cambridge/Mass: MIT Press, 2001), 90.

60 Deleuze and Guattari, *A Thousand Plateaus*, 25.

61 Ibid., 9.

62 Ibid., 249.

7 Returning to the Body: Orlan as an Ontologically *In-between* Subject

Deleuze and Guattari's formulation of the notion of becoming that proceeds via in-betweens or intermezzos – relational, open-ended and mutating spaces that connect multiple and divergent forms – provides a powerful theorisation to articulate Orlan's body of work, which is capable of forming collective spatial arrangements of interpenetrating elements without yielding to a unified, stable centre. While Orlan's multifaceted performance series visualises a radical approach of art, science, medicine entanglement that collapses their disciplinary borderlines, it is also a drastic alteration of the bodily space, which explores the intertwinement of corporeal human and non-human technology. As a matter of fact, by opening her body to the intrusion of a varied range of inorganic implants and prostheses that meld with her natural, corporeal properties, Orlan becomes a figure of interrelationality in one embodiment. The dissolution of the categories separating the organic flesh from the technological apparatus, the human from the non-human into an extended space of linkages and alliances yields to a corporeal ambiguity, diversity and indeterminacy that are disturbingly in-between. Similarly, Orlan's experimentation with primitive disfigurements, past tribal and ritual deformations in *Self-Hybridizations* series radically disrupts the recognisable human forms and dramatises the intertwining of self and other. With their perplexing forms, these images are appealing but uncertain in appearance. Seen in these terms, *Self-Hybridizations* contributes to the bodily ambiguity played out during the artist's preceding performative-surgery series and highlights a plural aesthetics that operate in-betweens of various bodies of different morphological, sociological and spatio-temporal origins.

As a logical outcome of these complex convergences between divergent forms and phenomena in Orlan's work, disrupting dualisms becomes a central aspect that makes up *The* in *Reincarnation of Saint Orlan* and *Self-Hybridizations*. Indeed, these projects present a radical attempt to challenge the binary distinctions and hierarchical categorisations between the spaces that belong to corporeality/ technology, natural/artificial, self/other, human/non-human, interior/exterior, visceral/ virtual, ancient/contemporary. While several scholars have already noted the binary-crossing nature of Orlan's body of work, I assert that her multi-media surgical project, along with her more recent digital self-portraits – which situate themselves outside the dualistic structures surrounding nature, artifice, human and non-human – in fact operate within the intermediary spaces or interstitial passages of Deleuze-Guattarian becomings. O'Bryan, for example, notes that, 'binaries that become unleashed from Orlan's work are subject/object, male/female, self/other, natural/unnatural, interior/exterior, beauty/the monstrous, art/medicine and so on.'[63] Indeed, Orlan's entire work conveys an understanding that challenges the hierarchical bodily structures in terms of dualities that separate man from woman, beautiful from grotesque, human from non-human, real from virtual, natural from artificial, instead it consists of radical forms of bodily exchanges and relational linkages between apparently contradictory terms. The way she makes use of diverse forms of technology to compose her art signals towards boundary-crossing couplings between the real and the synthetic, as well as a profound connectivity with otherness. It, thus, turns her corporeal subjectivity into a heterogeneous and collective site constituted by diverse spatial arrangements between organic, inorganic, technical, natural, artificial components that continuously resonate with one another. It is my key assumption that this strategy of opening up her corporeal space to exteriority, to multiple elements in a dynamic interaction posits Orlan's artistic act in-between *milieu* of pure metamorphoses, in the non-binary space when different realms of experience intermingle, (human/ non-human/ machinic/ organic/ inorganic/ real/ digital/ figuration/ disfiguration), which characterises the *intermezzos* of becomings. The work of Orlan, therefore, reveals further resonances with Prowse's analysis of the female drug literature, particularly with Garner's *Monkey Grip,* in that they are both providing potent instances of in-between spaces that have the potential to go beyond the dichotomised structures and fixed spatio-corporeal boundaries. While their mediums, approaches and issues of concern vary diversely, the drug trope in *Monkey Grip* that dramatises 'the temporal and spatial interstices between the dreamed drug spaces of the psyche and the "real" fragments of urban space'[64] as well as the blended in-between realms of nature/subject, urban/rural display striking parallels with Orlan's work, which equally generates a complex, relational and collective space.

This collectivity of connections, in turn, renders Orlan's art as unnatural, uncertain, ambiguous, transitory but ultimately transformative. These self-transformative processes occur through the expansive capabilities of these ambiguous, inexact spaces that express an openness to other bodies, technologies, objects and spaces. This kind of strategy: 'to be-between, to pass between, the intermezzo' Deleuze suggests together with Guattari, is the only way to bypass the oppositional discourse and get outside the dualistic logic.[65] One has to accentuate this perspective, as it might provide possibilities of a less enclosed

63 O'Bryan, *Carnal Art*, 116.

64 Prowse, in this volume.

65 Deleuze and Guittari, *A Thousand Plateaus*, 227.

and more connected existence, capable of engaging the in-between realms of becoming that are characterised by movement and processual change. As a matter of fact, Orlan's bodily space is always situated *in between*; in between figuration and disfiguration, in between real and virtual, in between natural and artificial, in between surgeries, in between processes of recovery, in between civilised and primitive, in between states of self and other. This would be a precarious body that emphasises its imprecise, transitional, indeterminate constitution by occupying a middle ground; an existence continually becoming-other-than-itself.

8 Conclusion

This chapter was an exploration into the dynamics of Orlan's art practice within a Deleuze-Guattarian framework. By taking the authors' conceptual tools into other fields – the *milieus* of the artist's complex body project – it attempted to produce lines of connections between their projects. It is the notion of becoming that seems to capture most fully the continual change the artist experiences, her body's dynamic, experimental, transitional states of being. Orlan, then, might be said to embody a being 'but only the being of becoming', as Deleuze would understand.[66] Elsewhere in this chapter, I have pointed to the ways in which the artist goes beyond traditional assumptions in which bodies are formulated as closed, integral and static unities according to preconceived schemes and spatial organisations, by approaching embodiment as a never-ending temporal process of becoming. In so doing, Orlan brings a range of alternative and unthought of corporeal possibilities into existence, she induces bodily ambiguities, uncertainties and differences that throw a potential challenge to the fixity of identity, in keeping with Deleuze and Guattari's articulation of *difference* over sameness.

In sum, I see Orlan's complex body project as emphasising the vital possibilities and open-ended spaces that might be opened – both in artistic sense and ontologically – once the conceptual barriers associated with binary thinking is evaded. Through her entire *oeuvre*, which enacts a comprehensive corporeal respatialisation, Orlan might, in fact, be questioning the traditionally hierarchical and oppositional ways in which we think about ourselves and the world, and open pathways for envisioning non-dualistic forms of bodily space that are no longer based on binary divisions. Thus, she opens the

66 Gilles Deleuze, *Difference and Repetition*, trans. Paul Patton (NY: Columbia University Press, 1994), 41.

ground for experimentation performed on the body and creation of previously unimagined forms of embodiment, provoking public debate about our bodies and their futures. Such provocations bring us back to what Orlan precisely wants her Carnal Art to do, to be a *public site for debate.* As it has been argued throughout this chapter, Orlan questions the binary modes of thinking and strives to move beyond the boundaries of accepted realities and spatialities towards the in-between spaces of potential connections and diverse possibilities. These intermezzos – spaces of bodily difference and encounter, instead of sameness and identity – are central to her whole body of work and perhaps also signal towards a future that evades more dualisms and more of the same.

Bibliography

Adams, Parveen. 'Operation Orlan.' *The Emptiness of the Image:Psychoanalysis and Sexual Differences*. London: Routledge, 1995.

Auslander, Philip. *From Acting to Performance: Essays in Modernism and Postmodernism.* London: Routledge, 1997.

Ayers, Robert. 'Serene and Happy and Distant: An Interview with Orlan.' *Body Modification,* edited by Mike Featherstone, 171–184. London: Sage, 2000.

Bogue, Ronald. *Deleuze and Guattari.* London and New York: Routledge, 1989.

Braidotti, Rosi. *Metamorphoses: Towards a Materialist Theory of Becoming.* Cambridge: Polity Press, 2002.

Braidotti, Rosi. 'Affirming The Affirmative: On Nomadic Affectivity.' *Rhizomes* 11/12 (2005/2006): np. Viewed on 15 January 2016. http://www.rhizomes.net/issue11/braidotti.html.

Brand, Peg Zeglin. 'Bound to Beauty: An Interview with Orlan.' *Beauty Matters*, edited by Peg Zeglin Brand, 289–313. Bloomington: Indiana University Press, 2000.

Davis, Kathy. '"My Body is My Art": Cosmetic Surgery as Feminist Utopia?' *The European Journal of Women's Studies* 4.1 (1997): 26–41.

De la Villa, Rocio. 'Orlan's Wardrobe.' *Orlan + Davidelfin: Suture, Hybridisation, Recycling*, edited by Rocio De la Villa and Pedro Alberto Cruz Sanchez, 15–42. NC: Murcia Cultural, 2009.

Deleuze, Gilles. *Foucault.* Translated by Sean Hand. Minneapolis: University of Minnesota Press, 1988.

Deleuze, Gilles. *Difference and Repetition.* Translated by Paul Patton. Columbia University Press, 1994.

Deleuze, Gilles and Félix Guittari. *Anti-Oedipus: Capitalism and Schizophrenia.* Translated by R. Hurley, M. Seem, and H.R. Lane. Minneapolis: University of Minnesota Press, 1983.

Deleuze, Gilles and Félix Guittari. *A Thousand Plateaus: Capitalism and Schizophrenia.* Translated by B. Massumi. Minneapolis and London: University of Minnesota Press, 1987a.

Deleuze, Gilles and Claire Parnet. *Dialogues.* Translated by H. Tomlinson and B. Habberjam. London: The Athlone Press, 1987b.

Goodall, Jane. 'An Order of Pure Decision: Un-Natural Selection in the Work of Stelarc and Orlan.' *Body Modification*, edited by Mike Featherstone, 149–170. London: Sage, 2000.

Grosz, Elizabeth. *Volatile Bodies: Toward a Corporeal Feminism.* Bloomington: Indiana University Press, 1994.

Grosz, Elizabeth. *Architecture from the Outside: Essays on Virtual and Real Space.* Cambridge/Mass: MIT Press, 2001.

Hirschhorn, Michelle. 'Orlan: Artist in the Post-Human Age of Mechanical Reincarnation: Body as Ready (to be re-) Made.' *Generations & Geographies in the Visual Arts: Feminist Readings*, edited by Griselda Pollock, 110–134. London: Routledge, 1996.

Ince, Kate. *Orlan: Millennial Female.* Oxford and New York: Berg Publishers, 2000.

Jones, Amelia. *Body Art/Performing the Subject.* Minneapolis: University of Minnesota Press, 1998.

O'Bryan, C. Jill. *Carnal Art: Orlan's Refacing.* Minneapolis: University of Minnesota Press, 2005.

Orlan. 'Intervention.' *The Ends of Performance*, edited by Peggy Phelan and Jill Lane, 315–327. New York and London: NYU Press, 1998.

Orlan. 'Harlequin Coat.' *Sk-Interfaces: Exploding Borders – Creating Membranes in Art, Technology and Society*, edited by Jens Hauser, 83–89. Liverpool: FACT and Liverpool University Press, 2008.

Orlan. 'Carnal Art Manifesto.' Accessed 18 February 2016. http://orlan.net.

Ruhs, August. 'Orlan Metamorphoser.' *Body as Membrane* nv (1996): 46.

Schultz-Touge, Christine. *Orlan: Carnal Art.* Paris: Flammarion, 2004.

Zimmermann, Anja. '"Sorry for Having to Make You Suffer": Body, Spectator, and the Gaze in the Performances of Yves Klein, Gina Pane, and Orlan.' *Discourse* 24.3 (2002): 27–46.

CHAPTER 3

Corporeal Interventions and the Contemporary Sounds of Keening

Michelle Collins

Abstract

This chapter discusses ritual sound as an intervention on the body that enacts respatialisation, creating a space inbetween, where there is the potential for transformation. This anthropological research investigates the ritual sounds of keening, as a corporeal intervention, within the context of the modern day Keening ceremony. In the preceding chapter Burcu Baykan examines the body not as a complete, bounded entity with distinct boundaries between interior and exterior, but rather as porous, leaky. The limits of the body's pliability are considered here also, where boundaries potentially dissolve and spatial-corporeal divisions collapse through interventions on the body via keening sounds. Conventional notions of space are transcended in Keening ceremonies and an ambiguous space emerges. This chapter suggests an alternative experience of being and relating as an outcome of ritual sound inscribed on the body. The chapter concludes with a discussion of the potential for individual and communal transformation through inhabiting inbetween spaces.

1 Introduction

This chapter turns from literature and performance art to a discussion on the performance of sound in ritual as a corporeal intervention. Keening was a pre-Christian Irish custom consisting of extemporised laments and wails sung by *bean chaointe* (also known as 'keening women' – women who sing the keen) at wakes and funerals in Ireland. This practice began to decline in the early twentieth century and died out in the 1950s. Once an essential part of Irish mortuary rituals, the sounds of keening now appear in an altered context today, presenting in a semi-private setting called Keening ceremonies.

This chapter will consider the potential for ritual sound to enact a 'highly charged ludic space outside of ordinary time'[1] where spatial-temporal-corporeal

1 Tríona Ní Shíocháin, 'Memory, Liminality and Song Performance: Understanding the History of Thought through Song,' *International Political Anthropology* 7.1 (2014): 1.

 | DOI 10.1163/9789004365520_005

divisions become blurred. It will focus on this inbetween space and threshold for transformation through the cultural practice of keening and the emergence of the contemporary phenomenon of Keening ceremonies.

Keening ceremonies are inspired by the pre-Christian custom of keening sung at wakes and funerals. Prior to this research, Keening ceremonies have not, as yet, been studied and recorded. For the purposes of this chapter, I examine data[2] collected through participant observation at a Keening ceremony; an analysis of a recording of a Keening ceremony; and interviews with participants and facilitators of the keen today.

By primarily focusing on data obtained through participant observation at this Keening ceremony, an analysis will be undertaken to illuminate the potential of alternative corporeal possibilities through inhabiting the inbetween space of the Keening ceremony. As well as this, sound in Keening-ceremonies – its potentiality in altering spatial and corporeal boundaries, making space and the body more fluid and permeable – is also considered. In alignment with Nycole Prowse's[3] suggestions in Chapter 1 as seeing sites of periphery as empowering, here the marginal, peripheral space is also examined for its potential to be a threshold for transition and transformation.

Victor Turner's concept of liminality will be used as a model for viewing and a framework for discussing the relationships present within the Keening ceremony. The concept of liminality considers the participants of the middle phase, the liminal phase, of rituals as 'neither here nor there; they are betwixt and between the positions assigned and arrayed by law, custom, convention and ceremony.'[4] Liminality is an intermediate state of being 'inbetween' in which individuals are stripped of their usual identity and their constituting social differences while being on the verge of personal or social transformation. The work of Julia Kristeva, specifically her ideas on the semiotic and symbolic,[5] will be used to further explicate the liminal experience of keening.

Turner suggests that the inbetween space of liminality allows for communitas to emerge. In communitas, people stand together 'outside' society and experience intense feelings of social togetherness, comradeship and belonging. Resonating with Burcu Baykan's[6] discussion on the rhizomatic *intermezzo*

2 Data was collected during research for a Masters of Traditional Arts from Høgskolen i Telemark, Rauland, Norway.

3 Nycole Prowse, 'Possibilities from the Peripheries: Time, Space and the Body in the Australian Drug Writing of Helen Garner and Luke Davies,' in this volume.

4 Victor Turner, *The Ritual Process: Structure and Anti-Structure* (Ithaca NY: Cornell University Press, 1969), 95.

5 Julia Kristeva, *The Kristeva Reader*, ed. Toril Moi (New York: Columbia Press, 1986).

6 Baykan, in this volume.

space in the preceding chapter, these concepts are applied in this chapter to explore ritual sound as an intervention on the body and the ensuing emergence of spaces of inbetweenness; to consider the possibility for the experience of communitas to emerge in this ambiguous space; and to discuss the potential for subsequent transformation.

In order to give a greater understanding of today's practice of keening in Keening ceremonies, the former practice at funerals, once so fundamental to Irish communities, will briefly be described.

2 Background: Keening at Funerals

2.1 *Context*

Contemporary keening is inspired by the pre-Christian custom called *caoineadh*/keening sung by *bean chaointe*. Keening was primarily a female tradition and was a fundamental part of funerary rituals – part of the 'fabric of Irish life.'[7] Keening took place at a time of 'disruption of the social order and this disruption is presided over and acted out by the lamenter.'[8] It was one of the rituals by which the community effected 'the re-establishment of social order in the face of the disruptive power of death.'[9] The sounds of the keen were significant to the community in guiding them through this ambiguous time of transition and re-assimilation into a new way of living, a newly formed community.

Outside of the funeral the keening woman was a regular participant in the community – a mother, a housewife, a nurse etc. However, for the duration of the funeral, the *bean chaointe* entered into what was described as 'divine madness,' inhabiting a dishevelled appearance, dressed in ragged skirt and shawl and voicing normally unacceptable sounds – 'barefoot and dishevelled, her hair loose.'[10] Some reports present the women as tearing their hair out and throwing it on the coffin, exposing themselves, wringing and clapping their hands or chewing on themselves while howling, shrieking and unleashing

7 Patricia Lysaght, 'Caoineadh os Cionn Coirp: The Lament for the Dead in Ireland,' *Folklore* 107 (1997): 65.

8 Angela Bourke, 'The Irish Traditional Lament and the Grieving Process,' *Women Studies International Forum* 11 (1988): 289.

9 Ibid., 289.

10 Narelle Phyllis McCoy, 'Madwoman, Banshee, Shaman: Gender, Changing Performance Contexts and the Irish Wake Ritual,' *Musical Islands: Exploring Connections between Music, Place and Research*, ed. Elizabeth MacKinlay, Brydie-Leigh Bartleet and Katelyn Barney (Newcastle, UK: Cambridge Scholars Press, 2009), 213.

'barbarous outcries' as they beat their breasts. On this 'semiotic borderline'[11] the bean chaointe were seen to be liminal, between worlds. This ambiguous space allowed her to express 'the collective outpouring of grief through her voice and body'[12] acting as a conduit, she helped those present to release grief. Just as Prowse[13] suggests of the potential of drug writing, keening provided a platform where restrictive social conventions could be challenged – in this peripheral position the keener was able to publicly voice and vehemently criticise public and private matters. Keening was a situation afforded to women where they could use their voice and be heard without constraint – 'articulating their own concerns and assorted social tensions.'[14]

When the funeral was over 'life was expected to return to normal and the lamenting woman to resume a normal role in society.'[15] Returning to her everyday role in society, the keening woman also returned to the 'symbolic order,'[16] the standard limitations experienced by women in Irish society of the time, no longer having the liberty to express freely as she had done while in the role of keening woman.

2.2 *Form of Funerary Keen*

We can glean from the descriptions given by visitors to Ireland from the seventeenth century onwards that the *caoineadh* was comprised of three parts: salutation, dirge/lament and *gol* or cry. The *salutation* can be defined as a fond address or call to the dead person. The *dirge/lament* or verse consisted of impromptu poetry from a body of stock poetry, partly extempore, partly prepared. This improvised poetry consisted of praising the deceased, mourning his/her passing and criticising the enemies of the deceased. It was at this point that the *bean chaointe* sometimes aired their concerns. The third part, the *keen/gol* is a choral cry led by the *bean chaointe* at the end of each verse. Through the *gol*/cry the *bean chaointe* created an aural space, which allowed others present to join in. The *gol* was repeated all through the performance of

11 Julia Kristeva, *The Kristeva Reader*, ed. Toril Moi (New York: Columbia Press, 1986).

12 Narelle Phyllis McCoy, 'Madwoman, Banshee, Shaman: Gender, Changing Performance Contexts and the Irish Wake Ritual,' *Musical Islands: Exploring Connections between Music, Place and Research*, ed. Elizabeth MacKinlay, Brydie-Leigh Bartleet and Katelyn Barney (Newcastle, UK: Cambridge Scholars Press, 2009), 213.

13 Prowse, in this volume.

14 Christina Brophy, *Keening Community: Mná caointe, Women, Death and Power in Ireland* (PhD Thesis, Department of History, Boston College, Boston 2010), Abstract 1.

15 Angela Bourke, 'The Irish Traditional Lament and the Grieving Process,' *Women Studies International Forum* 11 (1988): 290.

16 Julia Kristeva, *The Kristeva Reader*, ed. Toril Moi (New York: Columbia Press, 1986).

the keen. As part of the cry the word 'Och – ochóne!'[17] was repeated throughout.[18] The cry used no words and consisted solely of vocables like *och ochóne*, wails and outcries of grief. This aurally inbetween space allowed the community to give 'poignant expression to their emotion in purely musical terms'[19] releasing grief and transforming through it. The loose form of semiotic sound allowed for heightened ability to personally express both individual and collective grief. This ambiguous aural space allowed for unconventional sounds to be expressed by the community. These sounds, crossing borderlines between semiotic and symbolic spaces, were thought to pierce the veil between this world and the next, guiding the dead person's soul to the afterlife. At the same time, the sounds were believed to pierce the veil of grief, impacting all present and giving spatial opportunity to all to release their grief communally.

According to interviewee, Fr. Ó Conghaile 'the ochón has a certain tone to it. It has to do with the lifting of the state of the mind to a certain pitch'[20] where one can release grief and return to a more stable 'pitch' thereafter. He inferred that keening was when a pitch was reached which 'changed the balance,'[21] the sound intervening on the body this way, allowing listeners to release grief. Changing the balance 'the feelings of people come out. They (the keeners and

17 I have had difficulty in finding a direct meaning for the word 'ochón.' Ochón has no real meaning and loosely means 'my woe my woe' or 'alas.' Irish scholar Hugh Shields describes it as 'long-drawn-out vocables.' Two interviewees, Fr. Ó Conghaile and Emer Cloherty, believe the word ochón to be a spell word from Druidic times. I was not able to get any literature confirming this. However certain words in Irish have no meaning and are said as protections – e.g. seoithín seó. This is sung as part of a lullaby when putting a child to sleep in order to protect the child from the fairies and help him/her get to sleep. Mary Madden's 'A Bhean Úd Thíos' is said to be an example of this. By repetitive musical phrases and long-drawn-out vocables [*Seó hú leó, Seó hú leó* ag deireadh gach líne] she gives the message 'sleep!' to the child and to the fairies the message 'all is well.' The word ochón, as speculated by interviewees, has two functions also – to protect and guide the dead persons soul traversing from this world to the next and also to help bring out and release the grief of those present at the time. The interviewees suggested that there is no written record of the word 'ochón' as to do so would be 'like putting a sword in the hands of a child.' Due to the spell power of this word it could not be passed on to those that did not understand it.

18 Breandán Ó Madagáin, 'Irish Vocal Music of Lament and Syllabic Verse,' *The Celtic Consciousness*, ed. Robert O' Driscoll (Ireland: The Dolmen Press, 1981), 313.

19 Breandán Ó Madagáin, *Caointe Agus Seancheolta Eile, or Keening and other Old Irish Musics* (Connemara: Cló Iar-Chonnachta, 2005), 84.

20 Fr. Ó Conghaile, Personal communication (2013).

21 Ibid.

their sounds) get in at that level.'[22] In this way, keening is referred to as reaching a recognised tone or pitch, causing an effect on the keener and listener.[23] The *gol*/cry significantly extended the social function of the keen 'to that of a way for the community to release grief collectively.'[24] This final part of the keen allowed for the community to relate to each other during the grieving process and release grief together. It is this final part of the keen that has been most influential on the contemporary keen.

2.3 *Function*

Traversing the space inbetween, the parallel space between this world and the next, the keener was believed to use her voice and body to guide the dead person's soul from this world to that of the spirits using keening sounds as the intervening symbol connecting this world and the next.[25]

Similar to the impact of literature and performance art discussed as enabling in the preceding chapters, the willingness of the *bean chaointe* to inhabit the space inbetween and experience painful emotions allowed others present to do the same. In particular, the respatialisation of the keen parallels Baykan's Groszian examination of the empowering nature of the 'inbetween' that 'lacks a form, a givenness ... yet facilitates, allows into being ... a strange becoming.'[26]

Keening moved listeners to tears and facilitated them in expressing their own grief. The keener allowed and constructed the space for intense grieving for the dead to occur. As a result, the community could resume its regular pattern of life once mortuary rituals were concluded having expressed their grief through the 'intercession of the *mná caointe*'[27] and the sounds of keening as part of funerary rituals. This tradition began to decline in the early twentieth century. The contemporary phenomenon of Keening ceremonies has

22 Ibid.

23 Michelle Collins, *Caoine: Spaces for Vocalising Grief. The De-Ritualisation and Re-Ritualisation of Keening in Contemporary Ireland* (Rauland, Norway: Telemark University College, Department of Folk Culture, 2014), 29.

24 Breandán Ó Madagáin, *Caointe Agus Seancheolta Eile, or Keening and other Old Irish Musics* (Connemara: Cló Iar-Chonnachta, 2005), 84.

25 Patricia Lysaght, 'Caoineadh os Cionn Coirp: The Lament for the Dead in Ireland,' *Folklore* 107 (1997): 69.

26 Baykan, in this volume.

27 Narelle Phyllis McCoy, 'Madwoman, Banshee, Shaman: Gender, Changing Performance Contexts and the Irish Wake Ritual,' *Musical Islands: Exploring Connections between Music, Place and Research*, ed. Elizabeth MacKinlay, Brydie-Leigh Bartleet and Katelyn Barney (Newcastle, U.K: Cambridge Scholars Press, 2009), 214.

developed in the past ten years inspired by the former funeral keening. Next I present descriptions of the context, form and function in which the contemporary keen now appears.

3 Participant Observation Report

3.1 *Context*

The contemporary keen now presents itself in the context of Keening ceremonies, outside of the traditional funeral setting. Keening ceremonies are infrequent events where groups come together and keen. Unlike keening at funerals, at such ceremonies participants are not keening any one individual's death – every person keens their own personal grief. This grief is not always for the loss or passing of someone. Sometimes people keen in order to release grief surrounding a tragic event that happened in their life such as rape, inability to have children, abuse etc.

Keening ceremonies are typically held within the Celtic Shamanic[28] community. The ceremony I discuss in this chapter was an invite only event with all participants training or trained as Shamanic healers (excluding myself). It lasted approximately forty-five minutes. This was held in 'the temple,' a room in the facilitator's house regularly used for ceremonies and shamanic journeying. All present had experienced keening prior to this ceremony. All the attendants, excluding myself, were undergoing or had undergone shamanic training, working within the traditions of Ireland. Much like traditional keening that had taken place within the context of mortuary rituals, Keening ceremonies today take place within the context of this training and shamanic tradition.

Prior to the ceremony the facilitator 'cleared the space.'[29] This entailed burning sage and blessing the space where the ceremony was about to take

28 'Shamanic healing is a practice which establishes our connection to the whole of life and helps to restore balance. Life is the Great Tapestry – what we see is the picture of our lives and experiences. At the back of the tapestry are found the intricate thread connections that form a web of unity among all that exists. These connections are to be found in the invisible/ spiritual realms. Working with shamanic healing, we seek to reconnect these threads, focusing on the unity of mind, body and spirit. By addressing the spiritual causes of illness, the shaman, with aid of her allies, helps to restore balance between the seen and unseen world and helps the client to view the world in a whole and integrated way.' Cáit Branigan, Personal Communication (2014).

29 Michelle Collins, *Caoine: Spaces for Vocalising Grief. The De-Ritualisation and Re-Ritualisation of Keening in Contemporary Ireland* (Rauland, Norway: Telemark University College, Department of Folk Culture, 2014), 47.

place. Traditional keening – its form, function and impact – was described briefly to participants before the ceremony began, connecting and rooting the ritual with the past. Before entering the temple where the ceremony was to take place, the facilitator explained to us that we were free to make any sound we felt. There was but one rule – to make sounds that we felt and not to feign feeling.

Four of the six women were dressed in long skirts, black/dark in colour. Each participant had brought a shawl/veil, similar to traditional shawls, to cover them self with. I was given a veil to cover my head and face also. The purpose of this garment was to act as a symbol or resource to help individuals go inwards and access semiotic sounds, or as folklorist Angela Bourke describes it – to access that private place within them where grief is stored.[30] Veils may be symbolic of inbetween spaces, the veil between this world and the next, the inner world and the outer world. As a result, the use of the veil afforded a sense of the inbetween in this context. Having entered the temple we sat on beanbags in a circle.

At the commencement of the ceremony, the facilitator made participants aware that should they feel sick at any point, to raise their hand and a towel and basin would be brought to them. Following this the facilitator walked around each individual with a shaker. This act was to include and unite each participant in the 'sacred space.'[31] For the same purposes we were asked to stand and the following prayer was said:

> Breathe in; breathe in one another, the circle that has come together, in trust, in confidence, in unity. May we be a support to one another, may we each feel the freedom of expression … whatever grief and sadness that lies within. May we remember we are a circle together this day.[32]

Much like the circular gestures and movements made by the facilitator at the beginning of the ceremony around participants with a shaker, this prayer serves the function of uniting the group. Reminiscent of Baykan's discussion of an intermezzo space in the previous chapter, this opening prayer indicates

---

30 Angela Bourke, 'Keening as Theatre: J.M. Synge and the Irish Lament Tradition,' *Interpreting Synge: Essays from the Synge Summer School, 1991–2000*, ed. Nicholas Greene (Dublin: The Lilliput Press Ltd, 2000), 75.

31 Michelle Collins, *Caoine: Spaces for Vocalising Grief. The De-Ritualisation and Re-Ritualisation of Keening in Contemporary Ireland* (Rauland, Norway: Telemark University College, Department of Folk Culture, 2014), 47.

32 Cáit Branigan, Personal communication (2014).

the corporeal collapse (of bodies) through the respatialisation of the Keening ceremony.[33]

This space, aside from everyday living, is created for a non-judgemental domain where grief can be expressed in whatever way necessary.

3.2 *Ethnographic Experience*

The following description of the Keening ceremony, with edited excerpts from the participants, describes the impacts of the contemporary keen on the body and the result of this corporeal intervention in enacting respatialisation and transcendence of conventional conceptions of limits of the body.

> The lead keening woman begins to hum a melody quietly, making a closed sound. The facilitator joins with a more open sound, a counter melody, getting stronger. There seems to be no set rhythm. All present join over the course of a minute or so. The group becomes louder and the pitch rises. After a number of minutes the sounds become more dissonant and the atmosphere in the room changes, the mood becomes more sullen. With the rise in volume and pitch, sobbing increases and increases. It feels like something is about to break. The sounds become less melodic and are siren-like, screeching. The sounds are strong, raw and foreign to me – and although I know it is merely a group of women using their voices, a group of women I drank tea with moments before, I am struck with fear. The sounds are raw, primal even. Their complete unfamiliarity has a disorientating effect on me. My vision is distorted through use of the veil, and this combined with the impact of the startling sounds results in all appearing indistinct. I rock back and forth from where I sit in an effort to soothe myself. I hear crying and it is increasing in volume. The sound begins to travel in a wave around the group reaching a peak with one participant and then seemingly moving on to the next individual. It is as if the group is following each other's sounds, when one grieves loudly the rest of the group sounds louder. When an individual is experiencing an intense wave of grief it is as if the sounds of the other participants flow in support of this individual until the wave of grief has subsided. The mass of sounds move from individual soundings to group sounding as one and back again. The distinction between the individual and the group blurs. This mayhem of polyphonic voices then returns to quiet singing followed by a lull in the sounds. After this brief lull, the volume increases once

33 Baykan, in this volume.

> more and another round of grief is flowing. The noise is distressing and it has lost its melody. Coughing erupts around the room, low crying, then wailing. I begin to cough and get caught for breath. The group holds a high pitch in unison. We are a group with one voice again for a moment, the distinction between self and other, between private and public, between interior and exterior dissolving. Higher open pitches are sung and this is followed sharply by a disharmony. I begin to feel very sick. I cough loudly and gag. It feels like I am sliding and that an evaporation of centre, of stability, of limits is occurring. And with this decentred feeling the boundary between outside and inside slide, becoming slippery. There is fluidity between the group and me. The barriers between others and me seem distorted and my sounds seep into theirs, my sounds unite with theirs.
>
> The facilitator brings me a basin and towel and I cough into the bowl as I feel I am about to get sick. The sounds are dissonant at this point and I am coughing, coughing and coughing. Someone sings a vowel sound at a very high pitch and it makes me feel very sick to hear now. The pitch and volume goes up within the group until participants are screeching and wailing. It is at this point in the ceremony that the sounds are loudest, or so it sounds to me, however, it is less frightening and the group is unified in its expression. Support is communicated without words. I am the group momentarily. The group sing at this high pitch in unison, holding a high note, open-mouthed. I cough a lot. I am tired. I slump. I cannot really join the group singing. My feelings of sickness begin to subside and the group returns to closed sounds with the lead keener humming. It is as if the group has just carried me through my own wave of grief. The group hums in disharmony. A number of waves of grief follow and finally it is only the lead keener humming an improvised melody for a few minutes. To close the ceremony the facilitator prays, asking for healing from suffering, extending our hands and our hearts to those that are grieving around the world. She sings a song, "Ancient Mother." The ceremony is concluded. Standing up and coming together as a circle we give thanks for the space.[34]

The description of context and experience above identifies the contemporary Keening ceremony as a ritual:

---

34 Michelle Collins, Personal notes, participant observation (2014).

> In most religious rituals, the participants are collected into a place marked off as sacred and then participate in a series of activities that induce a certain emotional state – quiet awe, rapt attention, terror, intense pleasure, or frenzied enthusiasm, as the case may be.[35]

In ritual, individuals are taken outside normal space and time to a liminal state. In the constructed space for the Keening ceremony, separate from ordinary daily living, participants are invited into a sacred place, a 'temple,' a space they are aware is designed for ritual. Here the flow of ordinary conversation stalls. In Keening ceremonies participants are separated from ordinary life and ordinary sound-making.

The inbetween space of the Keening ceremony is a space where people have 'permission' to express themselves, show their 'vulnerability' in the group and communicate extremely private and raw experiences – 'you're absolutely exposing your most vulnerable side.'[36] This expression of extreme vulnerability in aural nakedness, the loose form of semiotic sound, is a shared expression among participants of the Keening ceremony.

The majority of participants interviewed about Keening ceremonies state the releasing and expressing of grief is the main function of the ceremony along with helping others to release grief.[37] Participants state that Keening ceremonies are a communication, letting others know of the grief they carry and sharing and supporting each other in the release of this grief. This is alluded to in the above description of the flow of support in sound. The contemporary keen, done in a semi-private context, aims to release personal grief in the ambiguous space of the ceremony and through inhabiting this space, bring the individual back to a place of balance.

The above description is engaged with in the following section through the framework of Turner's liminality.

4 Liminality

Turner's ideas of 'liminal' or 'liminality' are terms derived from the Latin 'limen' meaning *threshold.* Arnold van Gennep first introduced liminality in

35 William Sewell, 'Historical Events as Transformations of Structures: Inventing Revolution at the Bastille,' *Theory and Society* 25. 6 (1996): 870.

36 Deirdre Wadding, Personal communication (2014).

37 Michelle Collins, *Caoine: Spaces for Vocalising Grief. The De-Ritualisation and Re-Ritualisation of Keening in Contemporary Ireland* (Rauland, Norway: Telemark University College, Department of Folk Culture, 2014), 56.

his book *The Rites of Passage* (1960). Van Gennep describes Rite of Passage as having a three-part structure: separation, liminal period and re-assimilation/ incorporation. In the second phase of liminality the 'initiate' is stripped of his/her former status before a new status is conferred in the third phase and the individual is re-assimilated back into the community. The second phase is therefore an interstructural phase where status and identity are ambiguous. Fifty years after van Gennep's analyses, Turner appropriated the concept of the liminal as this transitional state.

Liminality is the state of ambiguity that occurs in the middle stage of a ritual when participants in the ritual are between statuses, 'neither here nor there; they are betwixt and between the positions assigned and arrayed by law, custom, convention, and ceremony.'[38] The conventional hierarchies and patterns that normally hold are suspended. Persons who find themselves in a liminal phase are 'temporarily undefined, beyond the normative social structure'[39] and are socially ambiguous for the duration of this phase.

In this Keening ceremony identity was firstly made ambiguous through the use of similar dress and veils. With the unspoken uniform of long skirt and black shawl/veil over our heads and faces, all participants took on the same appearance. Along with similar dress, the stripping of a participant's position and status is also experienced as we sit as equals in a circle on beanbags on the floor of the temple. These processes are symbolic of a social levelling of candidates whereby participants sit as equals on the floor and the identity of the individual is temporarily undefined as participants are barely recognisable through their veil. This state of stripping away, when the individual is not defined in terms of their status but in terms of their individuality, is an empowering gesture.

Turner states that liminality is characterised by the 'blurring and merging of distinctions.'[40] The removal of social distinctions between participants is the first borderline that begins to blur. In this space, where all participants are made to appear equal, they are free to express themselves as they wish. Participants undergo an embodied experience built up of multiple symbol sets, of which I consider sound to be essential. The sounds of keening are among what Turner refers to as the 'rich variety of symbols,'[41] which ritualise liminality and

38 Victor Turner, *The Ritual Process: Structure and Anti-Structure* (Ithaca NY: Cornell University Press, 1969), 95.

39 Victor Turner, *From Ritual to Theatre: The Human Seriousnesss at Play* (New York: PAJ Books, 1982), 27.

40 Victor Turner, *The Ritual Process: Structure and Anti-Structure* (Ithaca NY: Cornell University Press, 1969), 95.

41 Ibid., 95.

this transition. A physical space is created and yet an aural 'in between' space also, set apart from ordinary speech and sound making.

In this position, where structures are dissolved and participants are equal, corporeal intervention via keening sounds has a powerful impact. The intercession of these raw sounds has a disorientating effect often causing fear and distortion. As one participant notes: 'Their complete unfamiliarity has a disorientating effect on me.'[42] The experience was one of rawness, a primal exposition of oneself and showing of oneself in aural nakedness. In this ludic space, outside of convention and ordinary sound making, relation to space, to one's body and other bodies in the space, begins to alter. As can be read from the ethnographic notes the sounds take no form, no rhythm and no distinctive melody. Very quickly 'It feels like something is about to break … the individual sounds naturally merge into one. We are a group with one voice.'[43]

I suggest that my response of feeling sick in the ceremony can be associated with the spatio-corporeal disorientation and sense of ambiguity experienced by the intervention of raw, primal sounds and the unfamiliarity and uncertainty experienced during the ritual. Corporeal boundaries shift in the ceremony: 'I am sliding and that an evaporation of centre, of stability, of limits is occurring.'[44] In the collapse of spatial-corporeal divisions, the limits of the body become blurry and corporeal borderlines become displaced:

> We are a group with one voice again for a moment, the distinction between self and other, between private and public, between interior and exterior dissolving.[45]

Much like Baykan's[46] discussion of the performance art of Orlan and Prowse's[47] discussion of the drug trope, the bounded entity of the participant of a Keening ceremony becomes slippery and begins to seep into the collective. A fluid and dynamic relationship between the individual and group develops. Within the embodied experience of this fluid, transitional space a dynamic relationship emerges.

---

42 Michelle Collins, *Caoine: Spaces for Vocalising Grief. The De-Ritualisation and Re-Ritualisation of Keening in Contemporary Ireland* (Rauland, Norway: Telemark University College, Department of Folk Culture, 2014), 56.

43 Michelle Collins, Personal notes, participant observation (2014).

44 Ibid.

45 Ibid.

46 Baykan, in this volume.

47 Prowse, in this volume.

5 Communitas

'Betwixt and between' the blurring of structural, aural and spatial-corporeal divisions in Keening ceremonies communitas emerges. Communitas occurs on the periphery, existing in the inbetween space of the liminal phase: 'Communitas breaks in through the interstices of structure, in liminality; at the edges of structure, in marginality.'[48] Martin Buber (1961) explains communitas in terms of community:

> Community is the being no longer side by side (and, one might add, above and below) but *with* one another of a multitude of persons. And this multitude, though it moves towards one goal, yet experiences everywhere a turning to, a dynamic facing of, the others, a flowing from *I* to *Thou*.[49]

In the ethnographic description of the Keening ceremony above, this 'facing of the others'[50] is expressed: 'It is as if the group is following each other's sounds.'[51] This aurally-driven respatialisation and corporeal connection is ameliorating:

> When an individual is experiencing an intense wave of grief, it is as if the sounds of the other participants flow in support of this individual, until the wave of grief has subsided.[52]

The participants witness, support, experience and are *with* each other in their grief. With the sounds of keening 'flowing from I to Thou,'[53] an interconnected community is experienced. As one participant described in interview:

> As people's grief emerged it carried mine. My sound facilitating their sounds of grief and by me hearing that it became real for me; it becomes real grief then for me and for them.[54]

---

48 Victor Turner, *The Ritual Process: Structure and Anti-Structure* (Ithaca NY: Cornell University Press, 1969), 128.

49 Ibid., 127.

50 Michelle Collins, Personal notes, participant observation (2014).

51 Ibid.

52 Ibid.

53 Victor Turner, *The Ritual Process: Structure and Anti-Structure* (Ithaca NY: Cornell University Press, 1969), 127.

54 Deirdre Wadding, Personal communication (2014).

The grief of the other participant carried and supported this individual's grief while at the same time her sounds facilitated their grief.

In a phase of liminality the Keening group is presented with 'a moment in and out of time' where communitas potentially emerges as 'some recognition (in symbol if not always in language) of a generalized social bond.'[55]

Again, resonating with Baykan's reading of Deleuze and Guattari's intermezzo,[56] it is in the removal of hierarchies that we experience the undifferentiation of community. In the fluidity of the inbetween, transitional space stripped of status and identity, the potential for communitas is a distinct possibility where people have shared goals and values.

As suggested above, in Keening ceremonies spatio-corporeal boundaries collapse. Traditionally the sounds of keening were believed to pierce the veil separating this world and the next, occurring in the ambiguous space inbetween and guiding the dead person's soul through to the afterlife. At the same time the sounds were believed to pierce the veil of grief, moving listeners to releasing and overcoming intense grief. Along with piercing the veil of grief, keening today pierces the veil dividing self and other. The non-dualistic nature of this occurrence is that of communitas.

In a Keening ceremony the binary opposition between self and other becomes fluid:

> And with this decentred feeling the boundary between outside and inside sliding, becoming slippery. There is fluidity between the group and I. The barriers between self and other seem distorted and my sounds seep into theirs, merges with theirs.[57]

A dynamic flow between interior and exterior, between individual and other, permeates the limits of the body. The divisions between bodies are leaky, the borderline is porous: 'the distinction between self and other, between private and public, between interior and exterior dissolving.'[58] As the personal sounds of individual grief slip and permeate others grief, becoming one voice, the sounds merge and there is an experience of the group merging together.

---

55 Victor Turner, *The Ritual Process: Structure and Anti-Structure* (Ithaca NY: Cornell University Press, 1969), 97.

56 Baykan, in this volume.

57 Michelle Collins, Personal notes, participant observation (2014).

58 Ibid.

The merging of grief in the collective of the Keening ceremony is an experience of the 'generalized social bond'[59] Turner describes. In the liminal phase where communitas emerges there is a strong sense of belonging to the group. In the case of Keening ceremonies, the recognition of the 'social bond' is expressed and experienced in this aurally ambiguous space through the sounds of keening: 'the group is unified in its expression. And support is communicated. I am the group momentarily.'[60]

Existing as part of a larger reality, which is fluid in communitas, each self is intimately connected to all other selves in the experience. In the chapter that follows, Muramoto makes suggestions of the transformative impact of re-spatialisation whereby notions of space and the body are challenged, impacting on the wellbeing of the overall community.[61] This aligns with the shared human experience of keening, of grief and support and of interconnectedness in this liminal moment, which binds the group. Through the corporeal intervention via keening an embodied understanding of the 'universal order' and 'generic social bond' occurs in this liminal phase.[62] Through inhabiting this ambiguous space an awareness of the group and sense of belongingness to the group, 'the sentiment for humanity'[63] develops:

> To close the ceremony the facilitator prays, asking for healing from suffering, extending our hands and our hearts to those that are grieving around the world.[64]

At the end of the ceremony, this interconnectedness with all of humanity is expressed extending support to those grieving around the world.

Communitas 'is accompanied by experiences of unprecedented potency.'[65] And the potency of this experience can be transformational. The potential for

59 Victor Turner, *The Ritual Process: Structure and Anti-Structure* (Ithaca NY: Cornell University Press, 1969), 97.

60 Michelle Collins, Personal notes, participant observation (2014).

61 Katsuhiko Muramoto, 'In Search of a Relational Body: Reflections on Sustainable Architecture,' in this volume.

62 Victor Turner, *The Ritual Process: Structure and Anti-Structure* (Ithaca NY: Cornell University Press, 1969), 97.

63 Ibid., 116.

64 Michelle Collins, Personal notes, participant observation (2014).

65 Victor Turner, *The Ritual Process: Structure and Anti-Structure* (Ithaca NY: Cornell University Press, 1969), 128.

transformation through inhabiting the space inbetween is discussed in the following section.

6 Transformation

In a phase of liminality the individual is at a threshold, a point of transformation. Turner discusses the liminal phase 'as a realm of pure possibility whence novel configurations or ideas and relations may arise.'[66] The relatively unstructured nature of the transitional or liminal state provides the opportunity and the context for self-transformation and the development of new ideas and relationships.

> The position of the in-between lacks a fundamental identity, lacks a form, a givenness, a nature. Yet it is that which facilitates, allows into being, all identities, all matter, all substance.[67]

It is the 'expansive capabilities' of the inbetween, as described by Baykan,[68] that create the potential for transformation, the habitation of marginal spaces as empowering.

As mentioned above, the primary objective of Keening ceremonies is the transformative release and expression of personal grief. Participants also stated that through occupying the interstructural space of a Keening ceremony a communication took place: letting others know of the grief you carry and sharing and supporting each other in the release of this grief. The contemporary keen brings the individual back to a place of balance through inhabiting this space, relating, sharing and supporting with other participants in the space. In this way Keening ceremonies provide the opportunity for personal change and transformation 'for people to move through grief, there has to be a forum, there needs to be a space for people go through their grief.'[69]

When questioning interviewees further about keening and possible reasons for its re-emergence, the majority of participants commented on the changes taking place in Irish society in recent years and the period of 'uncertainty'

---

66 Victor Turner, *The Forest of Symbols: Aspects of Ndembu Ritual* (Ithaca NY: Cornell University, 1967), 97.

67 Elizabeth Grosz, *Architecture from the Outside: Essays on Virtual and Real Space* (Cambridge/Mass: MIT Press, 2001), 90.

68 Baykan, in this volume.

69 Lydia Kiernan, Personal Communication (2013).

occurring now, with great grief over 'emigration,' 'loss of jobs,' 'failure of economy' and 'how the government has behaved':

> There isn't that sense of certainty in people anymore about what's going to happen – am I going to have a job tomorrow? Are my kids going to be here in six years time (emigration)? Maybe it [keening] is reaching people because people need some way to cope or some way to release.[70]

In a society where participants feel isolated, marginalised, disconnected, unstable and unheard, this practice can potentially provide an experience of relatedness, unity, interconnectedness, belonging and support. As discussed in the previous section, the distinction between 'I' and 'other' becomes slippery in communitas. This intimate experience of merging and relatedness with the other initiates a sense of reciprocity and can also potentially develop a sense of social awareness within the participant. In communitas 'we may catch glimpses of that unused evolutionary potential in mankind which has not yet been externalized and fixed in structure.'[71] Through transcendence of conventional notions of time, space and the body, via corporeal intervention of ritual sounds such as keening, the 'potential in mankind'[72] is glimpsed. Resonating with Muramoto's[73] notion of the transformative impact of respatialisation discussed in the following chapter, responsibility is born out of such an experience: 'The wisdom that is imparted in sacred liminality is not just an aggregation of words and sentences; it has ontological value, it refashions the very being of the neophyte.'[74] Through the emergence and experience of communitas in Keening ceremonies the context for transformation is created whereby participants have the potential to transform their experience of contemporary reality. Having 'glimpsed' this potential and having an embodied experience of 'universal truth'[75] our very being is reformed.

In communitas people stand together 'outside' society and this strengthens society. Keening ceremonies provide a forum where participants are 'released

70 Michelle Collins, *Caoine: Spaces for Vocalising Grief. The De-Ritualisation and Re-Ritualisation of Keening in Contemporary Ireland* (Rauland, Norway: Telemark University College, Department of Folk Culture, 2014), 67.

71 Victor Turner, *The Ritual Process: Structure and Anti-Structure* (Ithaca NY: Cornell University Press, 1969), 128.

72 Ibid.

73 Muramoto, in this volume.

74 Victor Turner, *The Ritual Process: Structure and Anti-Structure* (Ithaca NY: Cornell University Press, 1969), 129.

75 Ibid.

from structure into communitas only to return to structure revitalised by their experience of communitas.'[76] This is the transformative potential of the inbetween.

7 Conclusion

This chapter investigated the impact of the ritual sounds of keening on participants of a Keening ceremony. It explored the frameworks of liminality and communitas for the potential of corporeal boundaries to become permeable and respatialisation to occur. The concepts of liminality and communitas capture the impact, experience and potential of keening sounds and the transitional states of being in a Keening ceremony.

Keening ceremonies allow us to discuss previously undiscussed forms of relating. The leakages and slippages across time, space and the body in the ambiguous space of the Keening ceremony enable a re-evaluation of corporeal possibilities and potential – an embodiment of interconnectedness where we can find our relations to 'the other.' Consideration of Keening ceremonies allows us to reconceptualise spatio-corporeal confines. Resonating with Prowse's[77] and Baykan's[78] assertions, this chapter highlights possibilities of open-ended spaces where we relate differently to others and to space. On the periphery, in the interstructural, we learn to relate and engage in an alternative way, opening up the potential to transform and to refashion our being. Through this embodied experience of interconnectivity a meaningful relationship with 'the other' is developed. And borne from this experience a sense of social awareness and responsibility may develop.

Bibliography

Bourke, Angela. 'The Irish Traditional Lament and the Grieving Process.' *Women Studies International Forum* 11 (1988): 287–291.

Bourke, Angela. 'Keening as Theatre: J.M. Synge and the Irish Lament Tradition.' *Interpreting Synge: Essays from the Synge Summer School, 1991–2000*, edited by Nicholas Greene. Dublin: The Lilliput Press Ltd, 2000.

76 Ibid.

77 Prowse, in this volume.

78 Baykan, in this volume.

Brophy, Christina. *Keening Community: Mná caointe, Women, Death and Power in Ireland.* PhD Thesis, Department of History, Boston College, Boston, USA, 2010.

Collins, Michelle. *Caoine: Spaces for Vocalising Grief. The De-Ritualisation and Re-Ritualisation of Keening in Contemporary Ireland.* Rauland, Norway: Telemark University College, Department of Folk Culture, 2014.

Grosz, Elizabeth. *Volatile Bodies: Toward a Corporeal Feminism.* Bloomington: Indiana University Press, 1994, Introduction.

Grosz, Elizabeth. *Architecture from the Outside: Essays on Virtual and Real Space.* Cambridge/Mass: MIT Press, 2001.

Hobsbawn, Eric and Terence, Ranger. *The Invention of Tradition.* Cambridge: Cambridge University Press, 1983.

Kristeva, Julia. *The Kristeva Reader*, edited by Toril Moi. New York: Columbia Press, 1986.

Lysaght, Patricia. 'Caoineadh os Cionn Coirp: The Lament for the Dead in Ireland.' *Folklore* 107 (1997): 65–82.

McCoy, Narelle Phyllis. 'Madwoman, Banshee, Shaman: Gender, Changing Performance Contexts and the Irish Wake Ritual.' *Musical Islands: Exploring Connections between Music, Place and Research*, edited by Elizabeth MacKinlay, Brydie-Leigh Bartleet and Katelyn Barney. Newcastle, UK: Cambridge Scholars Press, 2009.

Ní Shíocháin, Tríona. 'Memory, Liminality and Song Performance: Understanding the History of Thought through Song.' *International Political Anthropology* 7.1 (2014): 1.

Ó Madagáin, Breandán. 'Irish Vocal Music of Lament and Syllabic Verse.' *The Celtic Consciousness*, edited by Robert O' Driscoll. Ireland: The Dolmen Press, 1981.

Ó Madagáin, Breandán. *Caointe Agus Seancheolta Eile, or Keening and other Old Irish Musics.* Connemara: Cló Iar-Chonnachta, 2005.

Sewell, William. 'Historical Events as Transformations of Structures: Inventing Revolution at the Bastille.' *Theory and Society* 25.6 (1996): 841–881.

Shields, Hugh. *Narrative Singing in Ireland.* Dublin: Irish Academic Press, 1993.

Turner, Victor. *The Forest of Symbols: Aspects of Ndembu Ritual.* Ithaca NY: Cornell University, 1967.

Turner, Victor. *The Ritual Process: Structure and Anti-Structure.* Ithaca NY: Cornell University Press, 1969.

Turner, Victor. *From Ritual to Theatre: The Human Seriousnesss at Play.* New York: PAJ Books, 1982.

CHAPTER 4

In Search of a Relational Body: Reflections on Sustainable Architecture

Katsuhiko Muramoto

Abstract

Current discourse on sustainable architecture is often too narrowly defined and much of the discourse concentrates on technological questions, especially on energy efficiency. A key assumption unquestioned in this approach is the separation between the natural world and the subject, a one-way mode of causation defining the subject's instrumental relationship with nature. Under this Cartesian paradigm of subject-object duality all things are knowable and controllable and an object (i.e. nature) is considered primarily in terms of its utility to human beings – through technology humans can control the environment and manage resources in a way that meets humanity's needs and desires.[1] The Modernist slogan 'Form follows function' is now replaced by *Form follows energy performance*. Cloaked and obfuscated under the current approach in architectural practice are the relational connections to *the other*. Drawing on the work of Martin Heidegger, Maurice Merleau-Ponty and Tetsuro Watsuji, this chapter elaborates a structure of *reversibility* and attempts to reconceptualise the interconnectedness and 'interdependentness' between body, space and nature in relation to the current sustainability discourse in architecture. It argues that scientific and technological advancements alone are not sufficient for a sustainable future. What is urgently needed is a new paradigm where we become aware of relational bodies.

1 Introduction

Since the World Commission on Environment and Development published its 1987 report *Our Common Future*, more commonly known as the Brundtland Commission Report, on the strategy for sustainable development, architects and engineers have been forced to fundamentally reassess the way we design and construct buildings. The report defines sustainable development very

1 See Martin Heidegger, 'The Question Concerning Technology' in *The Question Concerning Technology and Other Essays*, trans. William Lovitt (New York: Harper & Row, 1977), 5.

 | DOI 10.1163/9789004365520_006

broadly as 'development which meets the needs of current generations without compromising the ability of future generations to meet their own needs.'[2] In *The Hannover Principles: Design for Sustainability*, William McDonough extended the definition of sustainability and explicitly addressed designer's role in the development of sustainability. In it he states as the first of nine principles to 'insist on rights of humanity and nature to *co-exist* in a healthy, supportive, diverse and sustainable condition.'[3] He goes on further to stress the immense societal responsibility of designers in shaping cultural values and to be responsible for the impact and consequence of design.

In current architectural practice, however, the focus of sustainability is defined often too narrowly as a technological question of energy performance. How can we slow down the depletion of existing resources, therefore extending their lifespan and leave some for the future generations? The answer to the question is to get the most out of the least non-renewable resources and to replace them with alternative sources of renewable energy. Such an approach is best illustrated by the concept of the 'net-zero energy building': a building which produces as much energy as it consumes. The ultimate goal of sustainable architecture, therefore, is to transform buildings into self-sustaining, *autonomous* machines by making them effective in reducing energy consumption while balancing energy use with the harvesting of alternative energy from wind, sun, rain, earth, etc. The Net-Zero building is extremely well insulated in order to eliminate heat loss (or gain, depending on the season) through walls, windows and roof. As a result, they quite often leave us disengaged from nature and other people for the sake of efficiency and saving energy. It is important to notice that the question of *Why do we save?* is substituted by *How do we save?* as the driving force for inquiry. Now, sustainability discourse in architecture is reduced to a question of efficiency and optimisation. Originally defined as a code of conduct in the Brundtland report, the aim of sustainability is to preserve our natural and built environment for *the other* by reducing our current energy usage. It is this care, responsibility and obligation for *the other*, both for future generations and nature, rather than *saving* energy for its own benefit, which has made sustainable development compelling.

I feel we must first interrogate current sustainable approaches in architecture and demonstrate how our instrumental relationship with nature still

2 World Commission on Environment and Development, *Our Common Future: Report of the World Commission on Environment and Development* (Geneva: United Nations, 1987), 16, accessed 7 May 2015, http://www.un-documents.net/our-common-future.pdf.

3 William McDonough, *The Hannover Principles: Design for Sustainability*, (Charlottesville, VA: William McDonough Architects, 1992), 5, accessed 18 May 2015, http://www.mcdonough.com/wp-content/uploads/2013/03/Hannover-Principles-1992.pdf.

remains unquestioned. Drawing on the work of Martin Heidegger, Maurice Merleau-Ponty and Tetsuro Watsuji, I will elaborate a structure of reciprocity, reversibility and subsequent *in-between space*, similar to Burcu Baykan's[4] examination of Orlan's work rendering on Deleuze-Guattarian notion of *becoming-other* and *intermezzo* in *Respatialising The Body.* Echoing how Nycole Prowse[5] challenges the mind-body distinction and attempts to reconceptualise the body-space schema from the Kantian tradition and then Elizabeth Grosz' phenomenological writings, and much like the exploration of embodiment and the liminal space in keening by Michelle Collins[6] in this volume, the chapter reconceptualises the human-nature relationship. Via an examination of the interconnectedness and reciprocity between body, space and nature, this chapter's focus seeks an alternative to the technologically conceived relationship between body and nature common in the current sustainability discourse in architecture.

2 Heidegger and Technology

Current sustainable practice in architecture is based on a one-way mode of causation – nature is there to serve us. This reification of the natural world was only possible by the separation between body and *the world.* As Harold Alderman argues in *Heidegger's Critique of Science and Technology:*

> In Heidegger's view, technology extends Nietzsche's metaphysics of the will ... Technology approaches nature as a repository if resources containing energies that are unlocked by technology, accumulated, stored, transformed, and utilized in the production of goods.[7]

He continues on to describe technology as 'nothing more than one expression of the will to power and the machines of technology are the means of expressing this will.'[8] According to Heidegger, the meaning of technology is not the mere process of making but 'is a mode of revealing. Technology comes to

4 Burcu Baykan, 'Respatialising the Body: The Ontological In-Between Subject in Orlan's Body of Work,' in this volume.

5 Nycole Prowse, 'Possibilities from the Peripheries: Time, Space and the Body in the Australian Drug Writing of Helen Garner and Luke Davies,' in this volume.

6 Michelle Collins, 'Corporeal Interventions and the Contemporary Sounds of Keening,' in this volume.

7 Harold Alderman, 'Heidegger's Critique of Science and Technology,' *Heidegger and Modern Philosophy: Critical Essays* (New Heaven: Yale University Press, 1978), 47.

8 Ibid.

presence [West] in the realm where revealing and unconcealment takes place, where *aletheis*, truth, happens.'[9] In his effort of describing Heidegger's definition of technology as a mode of revealing, Alderman compares two kinds of boating: a motorboat and a sailboat. While a motorboat, he describes 'overcomes the water through the power of its engine,' a sailboat 'moves on the water through the use of currents and wind direction ... one is always very aware of this dependency.'[10] Heidegger accounts the former as 'challenging forth,' and it 'puts to nature as unreasonable demands that it supply energy, which can be extracted and stored'[11] for humans' purposes. Nature is revealed as a means-end schema of human instrumentality. The latter, oppositely described as 'bringing forth,'[12] is the poetic revealing of bringing-wind-forth: *technē* and *poiesis*.[13] Contrary to the motorboat's use of force, a sailboat lets '[wind] come forth into its own ... into the arena of accessibility.'[14]

What is important in the following discussion though is not the sailboat's *harmonious and non-exploitative relationship* with nature, as explored by Arne Naess and other ecologists, but *techne*'s capacity for *bringing it forth* to *poiesis* (something poetic). A sail harnesses the energy of the wind in order to move on water without turning it into a *mere natural resource* and *reciprocally brings forth* the presence of wind. Wind does not exist as such but a sail makes it appear to our visual experience in a poetic manner. One cannot imagine an Alexander Calder mobile without breeze. In the following sections I will examine current architectural practice with the aid of this original sense of *techne* that permits *poiesis,* the poetic *bringing forth* and attempt to reconceptualise the fundamental relationship between us and the world we live in.

3 Bioclimatic Architecture

With Alderman's examples in mind, let us first examine bioclimatic design. As in the case of the motorboat an unsustainable house works against climate and relies on mechanical means of environmental control for providing comfort.

---

9 Heidegger, *Question Concerning Technology*, 13.

10 Alderman, *Heidegger and Modern Philosophy*, 49.

11 Heidegger, *Question Concerning Technology*, 14.

12 Ibid.

13 *Techné*, origin of the word technology, designates skill, craft, and art. It is more than mere making, it is a way of disclosure, of 'bringing-forth something.' *Poiesis* is the origin of the word poetry, and poetic way of 'bringing-forth something.'

14 Michael Zimmerman, 'Authentic Production as *Techné* and *Poiesis*,' in *Heidegger's Confrontation with Modernity: Technology, Politics, and Art* (Bloomington and Indianapolis: Indiana University Press, 1990) 229.

As a result, it neglects or avoids any meaningful relationship with the external environment. By contrast, bioclimatic design acknowledges the importance of the relationship between buildings and the environment and works with what climate can offer. The concept was first introduced by the Olgyay brothers in *Design with Climate: Bioclimatic Approach to Architectural Regionalism*, and defined as follows:

> to ensure the existence and well-being of biological organisms within the given climatic conditions (primarily of humans, but protecting biodiversity). Bioclimatic architecture relies heavily on architectural science, especially architectural energetics, but goes well beyond that. It rejects energy-wasteful and inhuman environments and fashion-dominated architecture.[15]

Bioclimatic architecture regards climatic context as a major generator for design, i.e. building orientation with regard to the sun, shade, wind and other climatic factors and realises complex interactions between the natural and the built environment. Creating cohesion between design and natural elements, bioclimatic design maximises the positive climatic factors in creating an ideal indoor condition for human comfort. Because it is intimately related to the climatic condition where it is located, bioclimatic design is often described as passive design or climate responsive design and helps to reduce the energy consumption by optimising the benefit of natural resources, thus avoiding dependency on mechanical systems.

In the case of bioclimatic architecture, the parallel to Alderman's example of a sailboat is obvious: rely on nature to provide energy. Bioclimatic architecture aspires to establish equilibrium between buildings and nature by creating a *harmonious* relationship to surrounding environments. Climate, however, in the case of bioclimatic design is understood as a sum of various sets of measurable meteorological parameters such as temperature, humidity, air pressure, wind speed, etc. It regards nature as a 'standing reserve'[16] where a valuable natural resource is waiting to be extracted and exploited by human use. Thus, its approach is still predicated on the concept of domination and control and

15 Victor Olgyay, 'Bioclimatic Approach to Architecture,' in BRAB Conference Report 5, National Research Council, Washington, D.C., 1953, Quoted in Doris C.C.K. Kowaltowski, Vanessa da Rosa Watrin, and Silvia A. Mikami G. Pina, 'Tradition and Thermal Performance: An Investigation of New Vernacular Dwelling in Campinas, Brazil,' in *Traditional Dwellings and Settlements Review*, 18.2 (2007): 80.

16 See further description of *standing reserve*, Heidegger, *Question Concerning Technology*, 17, 173.

consequently its solutions are necessarily technological. The Cartesian mind and body split and the subsequent problematic human centred morality that caused the sustainability problem in the first place still remains intact.

Furthermore, since the main goal of bioclimatic design is to achieve an optimum solution, its approach is often understood merely as a kind of environmental determinism. The Modernist slogan of 'Form follows function' is now replaced by *Form follows energy performance*, a new form of functionalism, only this time under the false moral obligation of *saving the planet*. Bioclimatic design gained renewed interest amongst *green* architects seeking sustainable development alternatives with the advent of advanced environmental analysis computer software such as Ecotect® and Climate Consultant®. These sophisticated building energy simulation tools allow architects to test how a building will perform under different design variables such as daylight, wind, etc. (Figure 4.1) By objectifying nature, one can consider climate to be a scientific description of an interacting physical system, thus reducible to simple and mechanical causes and effects. Climate, however, can be more broadly apprehended as an interconnected system of influences of nature, history and geography. Such a view is expressed by a Japanese philosopher, Tetsuro Watsuji in his book *Fudo*. He claims that climate must be understood 'as expression of subjective human existence and not of natural environment,'[17] and argues that one's environment is central in shaping one's self. For him, who we are is the result of the climatic space into which we are born, live and die. We will expand on Watsuji's notion of climate later but it is sufficient for now to note that only under objectification of nature can one reduce climate to weather.

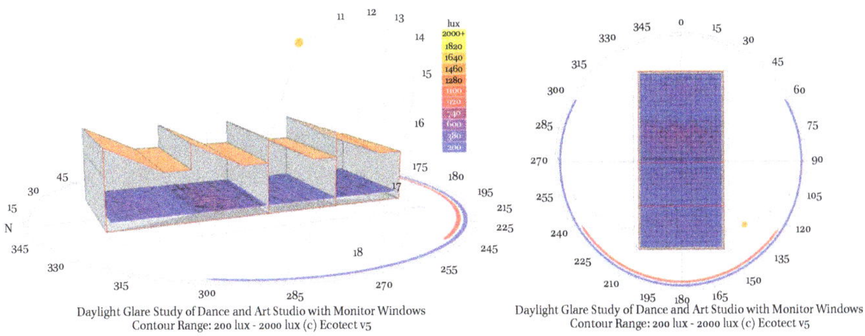

FIGURE 4.1 *Ecotect® rendering of daylight study for Kensington* CAPA *High School in Philadelphia (rendered by Moondeep Pradhanaga).*
© 2014 COURTESY OF THE AUTHOR.

17 Tetsuro Watsuji, *A Climate: A Philosophical Study*, trans. Geoffrey Bownas, (Tokyo: Ministry of Education, 1961) p. v.

4 Vernacular Architecture

Emerging from time-honoured tradition vernacular architecture refers to the type of buildings that are intimately related to their environmental contexts in a simple and humble manner. Adapted to climatic characteristics of the region, vernacular architecture used locally available materials; embraced its cultural building tradition of simple, but ingenious technology; and existed in harmony with nature as a part of local ecosystem.[18] Due to this special empathy with the environment an abundance of research has been conducted on vernacular houses through the lens of sustainability. As a result, vernacular architecture is often considered to be a dictionary of sustainable solutions and has been promoted as an ideal didactic model for sustainable design.[19] Much of the recent research, however, tends to emphasise the quantitative concerns of energy performance and isolates vernacular architectural components such as eaves and sun-shading devices and treats them as a set of recipes to be adapted in any new sustainable design while ignoring the relational/contextual and historical aspect.

Amos Rapoport describes in his *House Form and Culture* that 'building [a] house is a cultural phenomenon; its form and organization are greatly influenced by the cultural *milieu* to which it belongs.'[20] Emphasising the relational importance, he continues that 'house form is not simply the result of physical forces or any single causal factor, but is the consequence of a whole range of socio-cultural factors seen in their broadest terms.'[21] Another completely opposite yet common interpretation of vernacular architecture as a sustainable model is to use it as a reactionary argument to the uniformity of modern architecture and a *techno-centric bioclimatic* solution to sustainability. It renders vernacular architecture as the environmentally friendly approach that was built in harmony with nature because its construction materials were harvested locally without disrupting ecosystems, thus, claiming vernacular architecture still is a viable alternative sustainability model. This uncritical interpretation of vernacular architecture merely romanticises its harmonic

18 See Moholy-Nagy, *Native Genius in Anonymous Architecture* (New York, NY: Horizon Press, 1957), or Bernard Rudofsky, *Architecture without Architects: A Short Introduction to Non-Pedigree Architecture* (New Mexico: University of New Mexico Press, 1987).

19 See for example, *Vernacular Architecture: Paradigms of Environmental Response.* ed. Mete Turan (Aldershot: Avebury, 1990), or *Lessons from Vernacular Architecture*, eds. Willi Weber and Simos Yanna (New York: Routledge, 2014).

20 Amos Rapoport, *House Form and Culture* (Upper Saddle River, N.J.: Prentice Hall, 1996), 46.

21 Ibid.

co-existence with nature and assumes such socio-cultural factors evidenced in vernacular architecture can be *sustained* by simply imitating and repeating those design solutions in the age of globalisation. Such an approach of *form approximates idea* only glorifies the lost past and simply promotes a romantic provincialism. Neither the romantic provincialism of vernacular architecture nor a techno-centric bioclimatic architecture allows us to understand inter-connectedness between self, architecture and nature.

5 Reciprocal Belongingness and *Flesh*

Chapters in this section of the volume are tied together by their approach to problematise and de-centre the Cartesian-derived paradigm of subject-object duality while attempting to constitute a means of reconceptualising ambiguous relationships between self and other and self and world. Furthermore, such reconceptualisation is located differently in the reciprocal continuity and differentiation of *in-between*. In the following, I will draw a line of inquiry into the concept of *in-between*, drawing upon Maurice Merleau-Ponty's concept of reciprocity and reversibility and it is at this point a point that we find juncture with the thought of Heidegger:

> The enigma is that my body is simultaneously seen and is seen. That which looks at all things can also look at itself and recognize it, in what it sees, the 'other side' of its power of looking. It sees itself seeing; it touches itself touching; it is visible and sensitive for itself. ...This initial paradox cannot but produce others. Visible and mobile, my body is a thing among things; it is caught in the fabric of the world, and its cohesion is that of a thing. But because it moves itself and sees, it holds things in a circle around itself. Things are an annex or prolongation of itself; they are incrusted into its flesh, they are part of its full definition; the world is made of the same stuff as the body.[22]

In this passage, Merleu-Ponty introduces two key concepts pertaining to our discussion: *reversibility* and *flesh*. We can feel ourselves touching at the same time as being touched. This double belongingness of *my body-as-object* and the *body-subject I am* suggests that our relationship to *other* is open, polychromic and, most importantly, reciprocal. It is through this receptivity to things and

22 Maurice Merleau-Ponty, *Primacy of Perception*, ed. James. M. Edie (Evanston: Northwestern University Press, 1964), 162–163.

reversibility, according to Merleau-Ponty, which we become an integral part of the world. In his incomplete book, *The Visible and the Invisible*, he elaborates this concept of reversibility or *double-touch* further and introduces his concept of *flesh*. He tells us flesh 'is not matter, is not mind, is not substance'[23] and 'is not the union or compound of two substances'[24] but rather 'the sensible in a two-fold sense of what one senses and what senses.'[25] Here we find the kinship of Merleau-Ponty's argument with the 'constructed space of Keening – a communitus' as investigated by Collins[26] in creating a dynamic relationship between body and other and body and space. In addition, we can understand that Merleau-Ponty's reversibility thesis of seeing-being seen and perceiving-being perceived is reciprocity of activity and passivity – touch (active) is equally to be touched (passive). Therefore, we find overlap with the aforementioned Heideggerian notion of reciprocity.

In his earlier writing in *Phenomenology of Perception*, Merleau-Ponty argues how instruments become incorporated into the body schema: 'The blind man's tool has ceased to be an object for him, and is no longer perceived for itself; its point has become an area of sensitivity.'[27] In this instance, the cane became an extension of his body and was incorporated into his own corporeal schema and he later describes this in *Visible and Invisible* as 'things pass into us as well into the things.'[28] Elaborating further on Merleau-Ponty's embodiment of 'the blind man experiences world through his cane,' Don Ihde explains body and instrument are correlative and mediatic. Using eyeglasses as an example, he first describes that 'the technology is actually between the seer and the seen, in a position of *mediation*'[29] and continues on to note 'my glass becomes part of the way I ordinarily experience my surrounding; they "withdraw" ... I then actively embodied the technics of vision.'[30] As such, he argues that the instrument can extend the body's ability to act, perceive and to be perceived, thus they are incorporated into our corporeal schema, capable of establishing the experience of embodiment. He proposes a variation of body experience called

23 Maurice Merleau-Ponty, *The Visible and the Invisible: Followed by Working Notes*, trans. Alphonso Lingus, ed. Claude Lefort. (Evanston: Northwestern University Press, 1968), 139.

24 Merleau-Ponty, *Visible and the Invisible*, 140.

25 Ibid., 259.

26 Collins, in this volume.

27 Maurice Merleau-Ponty, *Phenomenology of Perception*, trans. Colin Smith (London: Routledge and Kegan, 1962), 143.

28 Merleau-Ponty, *Visible and the Invisible*, 123.

29 Don Ihde, *Technology and the Lifeworld: From Galileo to Earth* (Bloomington and Indianapolis: Indiana University Press, 1990), 73.

30 Ibid.

'extended embodiment'[31] – an 'instrument-mediated experience in which the instrument is taken into one's experience of engaging the world.'[32] With Merleu-Ponty's double belongingness and Ihde's extended embodiment, the phenomenological approach of I-experiencing-world can be now restated to I-experiencing through architecture-nature. In this case, architecture is similar to the sailboat, the enabling agent or medium that interfaces self/body and environment and now 'frames' reality for us as much as it 'frames' us as we learned from Merleau-Ponty. In other words, with mediation of a particular instrument we are bodily involved with it and the world appears to us in a particular manner, a particular mode of revealing as Heidegger may put it. In Merleau-Ponty's conception of embodiment, relations are medium specific. That is to say, instruments are not conceived as merely perceptual extension but rather have a significant role in how we perceive the world – a different mode of openness. This openness, in turn, may lead to a new kind of embodiment and eventually will affect our relationship with the environment. It is also to be noted that Merleu-Ponty does not limit this intertwining to the human/world (nature) conception but also extends it to people, although not explicitly, as he notes 'the experience of my own body and the experience of the other are the two sides of one same Being.'[33]

6 Watsuji's Ethics of In-betweeness

In these writings from Merleau-Ponty one can find direct correlation with Teturo Watsuji's ethics of betweeness: 'betweeness consists in the fact that self and other are divided from each other … and the same time that what is thus divided becomes unified.'[34] Much like Merleau-Ponty's flesh, Watsuji's theory of betweeness unites two opposing ends – individual and social, and temporal and spatial. His understanding of nature and climate is of particular relevance to our discussion of 'interdependentness' between body, space and nature. In the opening sentence of his book, *Fudo* (or *milieu* as Augustin Berque translates), Watsuji states:

31 Don Ihde, 'Chapter 6: Technology and Cultural Variations,' *Consequence of Phenomenology* (Albany, NY: State University of New York Press, 1986), 124–131.

32 Don Ihde, *Experimental Phenomenology* (Albany, NY: State University of New York Press, 1986), 141.

33 Merleau-Ponty, *Visible and the Invisible*, 225.

34 Tetsuro Watsuji, *Watsuji Tetsuro's Ringaku*, trans. Seisaku Yamamoto and Robert Carter (Albany: State University of New York Press, 1996), 35.

> I use our word *Fu-do*, which means literally, 'Wind and Earth,' as a general term for the natural environment of a given land, its climate, its weather, the geological and productive nature of the soil, its topographic and scenic features. ...It is not without reason that I wish to treat this natural environment of man not as 'nature' but as 'climate' in the above sense.[35]

In a subsequent chapter, he further articulates climate:

> Man's way of life has its own distinctive historical and climatic structure, the individuality of which is shown with the greatest clarity by climatic patterns governed by the limitations within climate. Climate, essentially, is historical; so climatic patterns are at the same time historical patterns.[36]

Explaining the cold from a phenomenological intentionality point of view, he states 'in feeling cold, we discover ourselves in the cold itself'[37] and '... it is "we," not "I" alone that experience the cold. We feel the same cold in common.'[38] He tells us that we find ourselves in the mutual relationship to climate although every individual experience may differ subtly. Wastuji does not reduced climate to mere meteorological features, rather claims it as a physical space where *mutual relationship* is manifested. Therefore, climate becomes way of life, or is *who we are*. Conversely, human-as-in-between discovers herself or himself in climate and therefore

> the various measures which are thus discovered, such as clothes, braziers, charcoal-burning, houses, blossom viewing, dyes, drains, anti-typhoon structures and the like, are of course what we ourselves have devised at our own discretion.[39]

Watsuji regards climate as central in shaping cultures where various relational interconnections are established, including us as bodies. This line of thought is addressed in the following, and it is worthwhile quoting in length:

> A house style is an established mode of construction, and this cannot have come into being without some connection with climate. The house

35 Watsuji, *Climate*, 1.
36 Ibid., 134.
37 Ibid., 3.
38 Ibid.
39 Ibid., 6.

> is a device for protecting ourselves both from cold and from heat. The style of architecture must be determined most of all by the degree of protection required against cold or heat. Then a house must be so built as to withstand storm, flood, earthquake, fire and the like. A heavy roof is necessary against storm and flood, though it may be disadvantageous in the event of an earthquake. The house should be adapted to these various conditions. …These various restraints and conditions are taken into account and accorded their degree of importance before the pattern of the house of given locality is finally established. Thus the determination of the architecture style of a house is an expression of the self-apprehension of man within climate.[40]

At first one may be puzzled by the environmentally behaviouristic tone of his descriptions of architecture, however, his argument is soon realised – any forms of cultural productions merge out of interplay between human and climate and they are intimately and reciprocally related to each other. Watsuji's concept of 'betweeness' has particular relevance in discussing self and nature.

7 *Engawa* as a Space for Reciprocal Belongingness

With the aid of Merleau-Ponty's double belongingness and Watsuji's exposition of the in-betweeness, let us examine the relationship between human and nature and how their reciprocal bond is realised in Japanese architecture. Many traditional Japanese houses are surrounded by *engawa*. Often translated into English as veranda, it is a narrow wooden walkway covered by the overhanging eaves of the roof that connects the rooms of the house and the garden. Kisho Kurokawa, in his effort to introduce a unique Japanese spatial concept of ambiguity to Western audiences, used the term *engawa*. In his argument, *engawa* is 'a symbiosis of inside and outside' and 'a space of communication.'[41] The betweeness of *engawa* interfaces and mediates interior and exterior, human and nature and eventually human and human. As such, Kurokawa's concept of *engawa* gives us an excellent starting point in articulating the interconnectedness between people, built environment (in this case a house) and nature.

The etymology of *En*, the first character of *engawa*, relates to the notion of edge or periphery both physical and social and, in turn, invokes multitudes of relationships. Kurokawa referred to it as *grey space*, neither black nor white.

40 Ibid., 6–7.

41 Kisho Kurokawa, *Philosophy of Symbiosis* (London: Academy Editions, 1994), 305.

Engawa is an ambiguous space that mediates between interior and exterior and public and private. (Figure 4.2 & 4.3) It brings nature closer to people and people closer together. Thus, *engawa* is the place where the process of threefold intermediation or transition we explored in Merleau-Ponty and Watsuji occurs and the mediatory function of *engawa* is further modulated by a series of screens placed on the inner and outer edges of the *engawa*.

It is worthwhile to describe how this array of permeable screens allows the *engawa* to be transformed to accommodate different relational conditions.

> A: All open – creates an uninterrupted relation to outside nature. One is fully immersed with nature, although inside. This openness is not only to nature but also to neighbours. In terms of energy efficiency, cross ventilation will cool interior temperatures down. (Figures 4.4, 4.5 and 4.7)

FIGURE 4.2 & 4.3 *Typical Japanese houses with* engawa.
© 2014 COURTESY OF THE AUTHOR.

FIGURE 4.4 & 4.5 *Screens partially opened: view from inside (left) and outside (right).*
© 2014 COURTESY OF THE AUTHOR.

Sudare (a screen made of horizontal slats of bamboo or wood) can be added to avoid sun penetrating into deeper part of the house. It shades sunlight, but allows a breeze to come through to promote air movement.

B: The outer layer glass screens are closed, but shoji screens (inner layer) are open. One is still engaging nature through the act of viewing, and is not completely cut off from neighbours. Glass screens allow heat gain from sun, thus keeping the indoors warmer. *Engawa* space can be used as a sitting room or entertaining informal guests.

C: The outer layer glass screens are open, but shoji screens (inner layer) are closed. *Engawa* space is ventilated, thus cooling the house. *Shoji* screens allow light but not view to come into a room. They function as thermal insulation.

D: Both outer layer glass screens and inner layer shoji screens are closed (Figure 4.6). This creates two separate spaces: the enclosed *engawa* space and an inner room screened by shoji. *Engawa* space functions like a solarium by affording the maximum exposure to the sun, and heat gained from the sun keeps *engawa* warm even in winter. This layer of warm zone

FIGURE 4.6 & 4.7 Engawa – shoji *and glass screens closed (left), and both screens open (right).*

> insulates the inner rooms. Due to the translucent shoji screens, light still penetrates into the inner room.
>
> E: Amado (wooden) screens drawn and both outer layer glass screens and inner layer shoji screens are closed. Achieving the greatest protection from outside, the house can cope with adverse conditions such as typhoons or extreme low temperatures on winter nights.

By occupying the perimeter of the house, *engawa* envelopes the house but also creates fluid boundaries between inside and outside allowing one to become aware of reciprocity of human and nature. It does not simply sit between inside and outside but rather re-configures subjectivity and objectivity, similar to 'the surface of an inexhaustible depth: this is what makes it able to be open to visions other than our own,'[42] as Merleau-Ponty describe his concept of flesh. Similar to Baykan's Deleuze-Guattarian analysis of intermezzo and Orlan's work in this volume, one finds there is no conventional binary opposition of in/out, self/other or architecture/nature in the experience of *engawa*. Architecture defines ambiguous, co-funding and reciprocal continuity between human and nature and through it one becomes aware of touching and being touched by nature. Previously, Merleau-Ponty taught us that our embodiment is medium specific. That is to say: instrumental mediation transforms our perception of the world in a certain way and we experience it in a particular way. Furthermore, according to Ihde, this transformation accompanies amplification and reduction, it amplifies specific aspects of reality while others are reduced. We can observe his theory in the function of screens in the *engawa* as they amplify some aspects of nature such as the wind, while reducing others, such as the influence of the sun.

8 Contemporary Example of *Engawa*

How can we create the openness of a traditional Japanese house within the high-density urban context of modern day Japan? Moreover, how can we create a meaningful connection to nature in the city where nature is absent or hardly anything to speak of? One of the most well-known and explicit explorations of traditional *engawa* can be found in the 1995 Curtain Wall House designed by Shigeru Ban. One can observe a literal interpretation of traditional *engawa* in this house but utilising contemporary materials. Traditional *engawa* was translated into deck space and placed on the second floor facing the east and south

42 Merleau-Ponty, *Visible and the Invisible*, 143.

side of the house. In this house 'curtains' replace the function of traditional glass sliding doors and shoji screen. Many houses designed by Tezuka Architects, such as Eaves House and Engawa House, are different interpretations of *engawa*. Designed in 2003, Engawa House, for example, is a long and narrow house built facing the client's grandfather's house, sharing a narrow strip of garden. The side facing the grandfather's house with large sliding glass doors mediates indoor and garden space providing natural light and cross ventilation when opened. Those examples are a natural modernisation of traditional *engawa* functioning as a mediator between human, built environment and nature. Although not necessarily the most advanced, they achieve some level of 'energy efficiency' since nature is an important factor of this interrelatedness.

9 Contemporary Double Skin Facade

In addition to the aforementioned quality of mediating between people, climate and nature, variations listed above suggest that *engawa* can function as a very sophisticated active double skin façade from the sustainable design point of view. The double skin façade is usually defined as

> a pair of glass skins separated by an air corridor (also called cavity or intermediate space) ranging in width from 20 cm to several meters.... The main layer of glass, usually insulating, serves as a conventional structural wall or curtain wall, while the additional layer is placed either in front or behind the main glazing. The layers make the air space between them work to the building advantage primarily as insulation against temperature extremes and sound.[43]

The Solarlux office building in Nijverdal, Netherlands serves as a cogent example to compare relative to the concepts found in *engawa*. Developed in collaboration with the researchers at TU Delft and Imagine Envelop BV, the Solarlux building features a highly flexible, sophisticated, energy-efficient double skin. According to its website, the Solarlux is 'a building in which the changing times of day and year could be sensed from within, through a glazed skin.'[44] The double skin of the Solarlux creates corridor like space, much like *engawa*, between

---

43 Harris Poirazis, *Double Skin Facades for Office Buildings* (Division of Energy and Building Design: University of Lund, 2004), 15, accessed 24 January 2015 http://www.dsbo.dk/Portals/0/Dobbelte%20facader%20Lund%20universitet.pdf.

44 'Sorlarlux Presents the Co2Mfortoffice,' accessed 19 September 2015, http://www.archello.com/en/project/solarlux-presents-co2mfortoffice/226948.

two layers of high insulation value windows that encircle the building. In an opposite arrangement from *engawa*, the Solarlux's outer skin is intended to be protection from the elements without much thermal function while the inner skin is designed as the primal thermal barrier. Both windows can be fully opened to create sense of connectivity to nature while offering maximum thermal insulation when fully closed. Very similar to *engawa*'s adaptability, the Solarlux's façade allows the client to control the desired indoor temperature throughout all four seasons. The double skin of the Solarlux building, however, is unfortunately completely disembodied and is understood simply as a technologically conceived mechanical *filter* that takes advantage of climate for creating optimum indoor human comfort. The consequence of such an approach is interestingly described in the TU Delft researchers' assessment of the building façade:

> user behaviour is a problematic aspect; training and a thorough understanding of the functionality of the façade are required to achieve optimum comfort levels in the offices. If the user is passive and does not adjust the façade according to current weather conditions, the office space can easily overheat.[45]

We have discussed at length how reciprocity between body, space and nature can be established. Opposed to Cartesian beliefs of mind-over-matter, Merleau-Ponty established that artefacts are capable of mediating our sensory relationship with world, in so doing they transform how we perceive. Similarly, Ihde illuminates how instruments structure our experiences and shape how we live. Their revelations lead to a number of questions. Does the sophisticated double skin façade of Solarlux make us aware of the reciprocity of touching and being touched by nature? Or does it allow 'poetic un-covering' of nature? Will one day the occupant open the window not for the sake of lowering the room temperature to the optimum condition, instead for enjoying early summer breeze?

10 Conclusion

There is no argument that the source of the environmental crisis we face today can be traced to the attitude towards technologies that we have developed.

45 Marcel Bilow, *International Façades – CROFT: Climate Related Optimized Façade Technologies* (Rotterdam: TU Delft, 2012), 321.

There is also no argument that renewable energy technologies such as photovoltaics and wind-power are the source for solutions that may eventually lead us to carbon neutral way of living. Being green is a requirement for any new houses and one can imagine reduction of energy consumption of such greenhouses equipped with energy star rated appliances. Contrary to that belief, some reports indicate that energy savings achieved by technological advancement such as more efficient heating cooling equipment and improved insulation does not necessarily result in a societal reduction in energy use. Energy consumption in fact continues to rise. Per capita in the United State we are using more energy every year. As it is known as Jevons' paradox, improvement in energy efficiency merely results in building bigger houses. Even those who call themselves green conscious people buy larger energy efficient refrigerators since they are affordable. Jevons' paradox, argued by David Owen and others, urges us to consider a more fundamental change in our relationship to nature and natural resources. Thus, approaches to sustainable architecture should not be confined within the field of technology but must be expanded to the much larger context of ethics as originally defined in the Brundtland Commission Report. As I described above, and paralleling Collins' transformative approach of interconnectivity found in respatialisation, *engawa* is simultaneously a very sophisticated double skin façade from a sustainable point of view and an embodiment of interconnectedness where we can find meaningful reciprocal relations to 'others,' including nature. The sense of reciprocity and double-belongingness and ethical responsibility born out of such a relationship will allow us to address the important issues of sustainability for the future that are currently ignored. Thus, what is urgently needed is not technological advancement but an alternative conception of relational bodies. As is suggested in the previous chapters' call for a renewed understanding of spatio-corporeal relationships and possibilities, we have to learn non-exploitive encounter with nature and learn to be the gracious receiver of the gift of nature. Only when we learn this relational body can we respond to the other and conduct ourselves ethically in the age of environmental crisis.

Bibliography

Alderman, Harold. 'Heidegger's Critique of Science and Technology,' edited by Michael Murray. *Heidegger and Modern Philosophy: Critical Essays.* New Heaven: Yale University Press, 1978, 35–50.

Bilow, Marcel. *International Façades – CROFT: Climate Related Optimized Façade Technologies.* Rotterdam: TU Delft, 2012.

Heidegger, Martin. *The Question Concerning Technology and Other Essays*. Translated by William Lovitt. New York: Harper & Row, 1977.

Ihde, Don. *Experimental Phenomenology*. Albany, NY: State University of New York Press, 1986a.

Ihde, Don. *Consequence of Phenomenology*. Albany, NY: State University of New York Press, 1986b.

Ihde, Don. *Technology and the Lifeworld: From Galileo to Earth*. Bloomington and Indianapolis: Indiana University Press, 1990.

Kowaltowski, Doris C.C.K., Vanessa da Rosa Watrin, and Silvia A. Mikami G. Pina. 'Tradition and Thermal Performance: An Investigation of New Vernacular Dwelling in Campinas, Brazil.' *Traditional Dwellings and Settlements Review* 18.2 (2007): 79–92.

Kurokawa, Kisho. *Philosophy of Symbiosis*. London: Academy Editions, 1994.

Merleau-Ponty, Maurice. *Phenomenology of Perception*. Translated by Colin Smith. London: Routledge and Kegan, 1962.

Merleau-Ponty, Maurice. *Primacy of Perception*, edited by James. M. Edie . Evanston: Northwestern University Press, 1964.

Merleau-Ponty, Maurice. *The Visible and the Invisible: Followed by Working Notes*. Translated by Alphonso Lingus, edited by Claude Lefort. Evanston: Northwestern University Press, 1968.

McDonough, William. *The Hannover Principles: Design for Sustainability*. Charlottesville, VA: William McDonough & Partners, 1992.

Olgyay, Victor. *Design with Climate: Bioclimatic Approach to Architectural Regionalism*. Princeton: Princeton University Press, 1963.

Poirazis, Harris. *Double Skin Facades for Office Buildings*. Lund, Sweden: Division of Energy and Building Design, University of Lund, 2004. Viewed on 5 August 2015, http://www.dsbo.dk/Portals/0/Dobbelte%20facader%20Lund%20universitet.pdf.

Rapoport, Amos. *House Form and Culture*. Englewood, NJ: Prentice Hall, 1969.

Watsuji, Testuro. *Watsuji Testuro's Ringaku: Ethics in Japan*. Albany, NY: State University of New York Press, 1996.

Watsuji, Testuro. *A Climate: A Philosophical Study*. Translated by Geoffrey Bownas. Tokyo: Ministry of Education, 1961.

Zimmerman, Michael. *Heidegger's Confrontation with Modernity*. Bloomington: Indiana University Press, 1990.

PART 2

Bodily Interventions on Material Spaces

CHAPTER 5

Public Space Design between Alienation and Appropriation: The Case of Parkour

Shelley Smith and Nicolai Steinø

Abstract

Urban public space in a Western context has become increasingly functionalised and pre-determined. Designations for use and by who, have become more prolific and have introduced formalised codes of accepted behaviour. This represents a barrier to the appropriation of public space by citizens, particularly when it comes to less conventional cultural practices. This in turn becomes a barrier for both inclusion and how space is defined and conceptualised. This is problematic in a notion of public space as the space of exchange and meeting place of 'the other,' and as an essential part of building an inclusive, tolerant and stimulating urbanity. In this regard, architecture, design and urban planning are neither innocent nor neutral agents in the process of conceptualising, interpreting and materialising space. Design has the capacity to foster uniformity and unambiguity or to invite plurality and ambivalence. Activating the physical practice of parkour, this chapter takes its point of departure in alternative urban practices to discuss the role of architecture, urban design and planning in materialising space between segregation and intolerance on the one hand and inclusion and curiosity on the other and the degree to which space is appropriated as place. Focussing on material details for jumping and climbing, as well as the overall spatial structure for trajectory, parkour emphasises the background and the foreground, while architectural urban space focuses on the middle ground. This chapter examines this potential and its implications for an urban architectural practice aiming for plurality and ambivalence, rather than uniformity and unambiguity. Further, in the context of this publication which intends a more permanent record of the dialogues and interweavings created by the papers presented at the 4th Global Conference on Time, Space and the Body in 2014, this chapter will draw out some of the points that demarcate potential meetings with the other chapters in this section.

1 Introduction

A young guy in baggy pants and sneakers comes flying over a low wall and rolls over onto a small patch of grass. Back on his feet and one hand on the railing of

 | DOI 10.1163/9789004365520_007

an adjacent fence, he jumps onto the flat roof of a low-lying garage in one continuous movement. From the roof he is soon down on a concrete slab, which becomes the runway for a subsequent vertical run on a windowless wall a few meters in front of him. After a somersault, which puts him in front of a tree, he jumps up and grabs a branch with both hands and swings back up onto a parapet. Soon after, he is out of sight.

This guy is using the elements of urban space as a forum for practicing parkour – a physical activity that utilises the body's natural ability to move in its surroundings. In doing so, he is interested in the material details of urban space for jumping and climbing, as well as the way the elements link in order to provide a self-determined trajectory through space. While the former triggers a focus on the foreground – or details and materiality – of urban space, the latter triggers a focus on the background – or structure – of urban space.

When the tourist to Paris enters the historical Place des Vosges, she is likely to be fascinated by the layout of the space, the symmetries of the surrounding buildings and the regularity of the tree plantings. In the Schouwburgplein Square in Rotterdam, another tourist surely notices the huge red crane-like floodlight posts which frame the otherwise diffuse modern space. The thin crane structures contrast the solid volume of a cinema building that sits on the square and lends it much of its identity.

The tourists are making sense of space as a system of elements, which are related through some sort of formal continuity or repetition. In doing so, they are focusing on the architectural middle ground – or formal elements – of urban space.

Design of urban space is the professional undertaking of architects, urban designers and landscape architects. Like the tourists, there is a tendency to focus on the middle ground – the architectural focus of the building – that diminishes the view and position of background and foreground, placing them in a specific situation in the service of the middle ground. In this situation, the background – the embedding of spaces within the urban structure – may be seen as playing a role in understanding the cartographic level of flows and access, while the foreground – the detailing of street and space inventory – may be seen as the level at which the final touch is added. However, what mainly defines urban space design is precisely these two 'diminished' factors in an interplay with each other: the overall spatial layout and its constituent material elements such as buildings, trees and pavement. Herein lies the potential of human bodies to impact on the transformation of space and the utilisation of public space as a space of negotiation and change and education.

The notion of foreground and background in the context of this chapter is a physical entity addressing the large-scale structural level of urban space, the small-scale materiality of the experience of urban space and the mediating relationship of the body between the two. There are though analogous examples of foreground and background as a metaphysical entity to be found in the other chapters in this section – also with the body in a mediating relationship. Iklim Topaloğlu and Pelin Dursun Çebi place the individual as body and senses in dynamic interaction with the spatiality of public space, both impacting and being impacted by space – causing change and being changed.[1] In their words, specific interventions of an artistic nature such as choreography, movement, objects and installations 're-create.'[2] They re-create actions that place the human body in the foreground against the background of space.

This theatrical device can be seen in direct correlation with Ana Carolina Lima e Ferreira's subsequent chapter and her case of the Theatre of the Oppressed playing out in the public spaces of the city.[3] Here the foreground of the bodies of the actors is in interplay with two types of background: the background of public spaces in which it is played out and, in addition, the background of public policy.[4] The body of the actor and the 'spect-actor' become tools for altering both social and political realities.[5] Here the theatre has the capacity to influence, alter and create behaviours and governing procedures to effect political change.[6] While in the context of Katerina Mojanchevska's chapter, the new relationship she describes between 'citizens' and 'institutions' also forms a foreground/ background set that can be described in terms of bottom-up and top-down approaches to public space.[7] Here it is specifically the view on this relationship that is key i.e. how the users of urban spaces view themselves and how they are viewed.[8] This has a direct influence on how these two approaches can result in policy making.

Common to the four chapters in this section of the volume, then, is the notion that mediating between the background and the foreground – whether physical or metaphysical – human bodies bear the potential to transform and to activate public space into spaces that can effect change through education.

---

1 Iklim Topaloğlu and Pelin Dursun Çebi, 'New Interventions in Public Space: Spatial Readings through Art and Architectural Installations,' in this volume.

2 Ibid.

3 Ana Carolina Lima e Ferreira, 'The Urban Challenge of Participation through Theatre in Rio,' in this volume.

4 Ibid.

5 Ibid.

6 Ibid.

7 Katerina Mojanchevska, 'Self-Organised Citizens' Groups and Urban Space: Challenges in Planning Paradigm,' in this volume.

8 Ibid.

2 The New Nature of Public Space

Classical urban spaces, the marketplace, the town square, the church surrounds, are elements of the pre-industrial city, which was based on trade and hence was a precondition to the generation of wealth. To a large extent this wealth was invested unproductively in the city, in the form of art and monuments. Thus, the city as such was an artwork and its quality lay in its wholeness and use value. This city was the setting of an urban life, an urbanity, which provided its inhabitants with the fundamental urban qualities that are necessary, in Lefebvre's words, in order 'to inhabit' rather than simply 'to reside.'[9]

The advent of industrialism implied a crucial change in the nature of the city. The industrial economy became the driving force of urban transformation and development and use value was replaced by exchange value as a parameter for urban development. And by the rational organisation of the city as a machine – a tool for production – its spaces were divided into functionally separate units (housing areas, industrial areas, business districts, etc.) and its unity and artwork character dissolved.[10]

In today's post-modern economy, the identity of the individual is increasingly defined by its consumption of goods and leisure.[11] This has led to a change in the use of urban space. The city has become an object of consumption rather than of use and the gaze on the city has become that of the tourist – a passive and spectating gaze – rather than that of the local dweller.[12] The tourist's gaze on the city is removed and aesthetic. The city is experienced primarily as a setting, as a staging of leisure and consumption. Architecture is reduced to surface – facades – and the built environment is detached from its social context.[13] Maarten Hajer and Arnold Reijndorp call for change:

9 Henri Lefebvre, 'Right to the City,' *Writings on Cities* (Oxford: Blackwell, 1996), 147–160.

10 Ibid.

11 Sharon Zukin, 'Postmoderne urbane landskap: A kartlegge kultur og makt,' *På Sporet av Byen* eds. Jonny Aspen and John Pløger (Oslo: Spartacus, 1997), 159–190 and Sharon Zukin, 'Whose Culture? Whose City?' *The City Reader*, ed. Richard T. Le Gates and Frederic Stout, 2nd ed. (London: Routledge, 2000), 131–142.

12 Ditmar Steiner, 'Die Stadt zwischen der Sehnsucht nach Harmonie und den Gesetzen der Dienstleistungsgesellschaft,' *Risiko Stadt? Perspektiven der Urbanitet*, ed. Ullrich Schwarz (Hamburg: Junius, 1994), 216–221.

13 Kim Dovey, *Framing Places: Mediating Power in Built Form* (London: Routledge, 1999) and Neil Leach, *The Anaesthetics of Architecture* (Cambridge, MA: MIT Press, 1999).

> We must rekindle the lost relationship between the social and the physical space, between form and meaning, with an eye to differences and relationships, as well as bear in mind the demands of a mass culture in flux.[14]

There is a clear indication that a fundamental shift in the constituent social and spatial makeup of public space has occurred and this calls into question who the stakeholders and users are and how space is defined. This gives cause to re-address public space in a situation where the notion of space and the notion of public have become distanced from each other with the introduction of the words 'semi' and 'private' as describers of the spaces inhabited and used by citizens as 'public' and, in the case that economic and political interests dictate the design, use and user group.

3 Threats to Public Space: Barriers to Exchange and Appropriation by Citizens

Contemporary urban life and contemporary urban form are characterised by some of the same factors – speed, flux, complexity – and although this can impact the location and duration of public encounters, it is the encounter itself that is of paramount societal importance – defining who we are, how we are by underlining freedom of movement and accessibility to assembly. This is also a point made in various ways in the other chapters in this section. Urbanist Ali Madanipour sees free expression for all groups of the diverse societies of cities as a necessary function of public space for issues of both health and safety of the city entity itself.[15]

Public space makes contact between diverse societal groups possible by providing a physical forum for its happening. But where the notion of 'meeting' exemplifies the function of traditional forms of public space, 'exchange' is a more appropriate term for the type of contact most often occurring today. Less committal and more fleeting, an exchange need not involve physical contact but can be purely visual or auditory, it can happen in passing and be a product of proximity rather than direct contact. It provides a forum for the

---

14 Maarten Hajer and Arnold Reijndorp, *In Search of New Public Domain* (Rotterdam: NAi Publishers, 2001), 227.

15 Ali Madanipour, 'Public Spaces of European Cities,' *Nordic Journal of Architectural Research* 1 (2005): 7–16.

observance of the other – even the unknown other or the perceived 'dangerous' other – without risk. Such a forum is instrumental in breaking down barriers that sanction and encourage exclusion. Within the physical forum the experiential component is of the utmost importance to public spaces and not only how this is afforded but also whether it is afforded at all, becomes a key issue.[16] Hajer and Reijndorp suggest:

> The core of successful public space thus lies not so much in the shared use of space with others, let alone in the 'meeting,' but rather in the opportunities that urban proximity offers for a 'shift' of perspective: through the experience of otherness one's own casual view of reality gets some competition from other views and lifestyles.[17]

Through public space the opportunity to experience diverse groups – groups that we do not belong to – is afforded. And this, in fact, is how public space is defined by its users.[18] Public space should by definition encourage accessibility and openness in order to include diverse groups, as Mojanchevska, Topaloğlu/ Çebi and Lima e Ferreira[19] also state, but in a contemporary situation where political and economic agendas play out in public space, plurality gives way to pre- determination of function and user. Traditional public space combined a number of functions and activities: trade, political forums, the place of ceremony and ritual. Today, however, there is a shift to either mono-functionality or neutrality – or both, with the ensuing result that public spaces dictate nothing or too much and leave no room for imagination. Relegated to spaces of leisure or consumption, public spaces restrict free movement, free expression and inclusivity by dictating when, how and by whom these spaces are to be used. Formalised codes of behaviour ensure general themes of 'niceness' and a lack of conflict and friction.

Common attitudes regarding public space often revolve around ensuring safety and belaying the fear of violence. In this regard, tendencies of parochialisation and aestheticisation limit who is welcome and how this message is conveyed.[20] These factors are detrimental to the actual experience of public space as the space of 'the other.' As Yi-Fu Tuan alludes:

---

16 Lyn Lofland, *The Public Realm: exploring the city's quintessential social territory* (New York: Aldine de Gruyter, 1998) and Hajer and Reijndorp, *In Search of New Public Domain.*

17 Hajer and Reijndorp, *In Search of New Public Domain*, 89.

18 Lofland, *Public Realm* and Hajer and Reijndorf, *In Search of New Public Domain.*

19 Mojanchevska, Topaloğlu/ Çebi and Lima e Ferreira, in this volume.

20 Hajer and Reijndorp, *In Search of New Public Domain.*

> Experience is the overcoming of perils. The word 'experience' shares a common root (per) with 'experiment,' 'expert,' and 'perilous.' To experience in the active sense requires that one ventures forth into the unfamiliar and experiments with the elusive and the uncertain.[21]

4 Design and Architecture as 'Guilty Parties'

As hinted in the opening of this chapter, design is not innocent and it is not neutral. It conveys meaning and indicates behaviour through choices made in spatial composition, material selection and detailing decisions. Sometimes this is a deliberate process on behalf of the commissioners of the design, whether public or private. At other times it is simply a product of tacit understandings and assumptions. Whether the designer is aware of this or not, public space design indirectly selects its users as a function of the design, discouraging some and encouraging no one to claim space. Alienation and exclusivity are likely to ensue.

Ownership in the sense that there is a freedom to make these spaces one's own, to inhabit, to make use of inventory and urban furniture for whatever purpose, one imagines, is negated, for instance, by the use of 'speed bumps' and 'arm rests' on urban furniture and planters. Particularly obvious as add-ons to existing furniture, they are in general more prevalent in the urban public space picture and serve a functional purpose in the message that some behaviours are not wanted and some users are not welcome, in this case skaters and the homeless. Exclusion thus becomes a parameter in the design of urban spaces.

These marginalised user groups and their alternative practices are deemed to be at best undesirable, shabby, dirty and at worst are seen as dangerous and threatening. As a forum for exchange also in contemporary societies, public space can provide a social educating function that makes diverse groups visually available to one another thereby increasing familiarity and reducing perceived danger on the basis of the group or their practices belonging to the category of the unknown.[22] However, the function of social education is likely to be inhabited in spaces that restrict some groups and practices. Architecture and design become agents of control, they become the guilty party in fostering uniformity and unambiguity but they also have the potential to invite

---

21 Yi-Fu Tuan, *Space and Place: The Perspective of Experience* (Minneapolis: University of Minnesota Press, 1977), 9.

22 Lofland, *Public Realm*.

plurality and ambivalence. Hajer and Reijndorp call for a reconceptualisation of the notion of design:

> What we are searching for is the room that the design can provide for the 'occupation' of the space with a multiplicity of meanings. Public domain, in this view, needs not 'de-sign' but rather 're-sign': the invitation to occupation by new meaning.[23]

The power of public space to make transparent and to educate, e.g. in Lima e Ferreira's analysis of Theatre of the Oppressed or in Mojanchevska's self-organised citizen groups, can on a very practical level be seen as a collection of techniques and methods that serve to address diversity and inequality.[24] Lima e Ferreira's Theatre of the Oppressed utilises tools such as *visualising* – making certain behaviours and practices that usually occur in private, public or *distancing* – a method entailing public showing of damaging and violent behaviour as acting in a public (dis)play.[25] Both Mojanchevska's self-organised citizens groups and the Lima e Ferreira's Theatre of the Oppressed give a voice to diverse and potentially excluded publics.[26] These are practices in public that intend a breaking down of borders and they can be related to the strategies and interventions utilised in the creation of physical public spaces. As mentioned, design can encourage or discourage, alienate or include. In their book *In Search of New Public Domain*, Maarten Hajer and Arnold Reijndorp present a 'call to action' for planners, designers and policy-makers involved in the creation of contemporary public space.[27] They too suggested addressing diversity and making public spaces of social education through the use of tools and methods by offering up a toolbox of three planning strategies and five spatial interventions that can be seen as a kind of kick-start to working with urban spaces as the site of actual exchange and change.[28]

The strategies suggested by Hajer and Reijndorp are *Theming, Compressing* and *Connecting*.[29] *Theming* refers to the creation of spaces that have meaning for specific groups and not just creating space as an architectural or

23 Hajer and Reijndorp, *In Search of New Public Domain*, 101.
24 Lima e Ferreira, Mojanchevska, in this volume.
25 Lima e Ferreira, in this volume.
26 Lima e Ferreira, Mojanchevska, in this volume.
27 Hajer and Reijndorp, *In Search of New Public Domain*, 16–17.
28 Ibid., 116–133.
29 Ibid.

aesthetic device.[30] Related to this but with a focus on creating diversity through experience, *Compressing* works with organising elements that are meaningful for specific groups in close proximity with elements that are meaningful for other groups.[31] And finally the strategy of *Connecting* focuses on methods e.g. montage, *mis èn scene*, set changes that address the relationship different places have to each other.[32] As such it can work with various intentions such as confrontation and/or seduction and takes its clues from the world of theatre, an interesting connection to Lima e Ferreira.

In addition to the three strategies, the following interventions are described by Hajer and Reijndorp as instruments in the development of public space that are in sync with a contemporary situation:

- *Fences for public access* entails the creation of a visibly demarcated but still available territory that frames particular groups. In doing so it encourages visual availability and potential exchange between non-homogeneous groups.
- *Connections* being made between the private and the public that are mutually observed and exhibited. Here the notion of spectacle and stage is set into play and it is suggested that the active use of circulation space is used for this. This could further the potential for exchange in a dynamic urban situation rather than a static 'meeting.'
- *Liminal spaces* indicates the use of transitional, overlapping spaces in between homogeneous and specified enclaves creating a kind of buffer zone that not only defines the enclave, but also supports the maintenance of the enclave. The liminal space itself can both provide a zone of exchange or a support in the organisation of heterotopias.
- *More friction* challenges the present striving for friction-less and conflict-free space by inviting difference, conflict and change from the point of departure that friction-less space is not only boring, but stagnating.
- *Mobility and marginality* points towards a need for re-thinking mobility and movement e.g., cars and marginalised groups, e.g. the homeless, skaters as full-fledged and included parts of urbanity and the urban spectacle.[33]

With this catalogue of strategies and interventions, architects and planners, policy-makers and cultural authorities are challenged to actively work through

30 Ibid., 117.

31 Ibid., 117.

32 Ibid., 117–120.

33 Ibid., 120–133.

architectural, urban and design intervention, creating public spaces that encourage inclusivity and exchange in public space rather than creating public spaces that merely collect people.

> Public domain centres around experiencing cultural mobility: for the opportunity to see things differently the presentation of new perspectives, as much as the confrontation with one's own time-worn patterns, Being coerced to conform does not tally with this perspective of a properly functioning public domain. Being challenged to relate to others does.[34]

In siding with the tourists in their focus on the middle ground, architecture and design also adopt an aestheticising gaze and an implicit bias towards public space as a space for consumption rather than habitation and a space that does not foster inclusion. Excluded groups and practices, however, by necessity or conviction, often short circuit the design of their dismissal by finding alternate routes, ingenious and imaginative solutions and by activating the potential in objects and spaces re-signed.

5 Urban Planning

Apart from being the object of architecture and urban design, urban space is also the object of urban planning. While architecture and design are aimed at the immediate physical design of urban space, urban planning is aimed at the distribution of uses and the provision of services and more indirectly at urban space as a framework for the quality of urban life.[35] Yet uses and services cannot be distributed in space without resulting in some kind of urban form, as little urban form can be designed without consequences for the distribution of uses and services and their implied consequences for the quality of urban life. Hence, urban planning is an inherently political activity (and architecture and design indirectly so) as it is a question of 'who gets what, when, where, why and how.'[36]

34 Ibid., 116.

35 Nicolai Steinø, 'Vision, Plan and Reality: Urban Design between Conceptualization and Realization' (PhD diss., Aarhus School of Architecture, 2003).

36 Paul Davidoff, 'Advocacy and Pluralism in Planning,' *A Reader in Planning Theory*, ed. Andreas Faludi (Oxford: Pergamon Press, 1973), 292.

Urban planning has taken on different forms across time and space in order to accommodate different political aims, even if it has not always been conscious let alone explicit about its political nature. Since the heyday of urban planning around the middle of the twentieth century when it was seen as a scientific and synoptic endeavour capable of controlling the urban environment based on systematic analysis, it manifested itself both as a predominantly conservative activity, a radical activity aimed at social change as well as communicative and democratic activity aimed at gradual change.[37]

Along with these changes, urban planning has naturally shifted its methodologies. As described in Lima e Ferreira's account of urban street theatre as a planning approach, more radical planning approaches find themselves quite remote from the 1950s idea of a scientific planner in a lab coat equipped with synoptic knowledge. Likewise, more communicative and democratic planning approaches[38] may be remote from people who are not familiar or willing to engage in the types of communicative processes defined by such forms of urban planning. In addition, even the most well-meaning planners may have a blind spot for the views, ideas and communicative formats of certain cultural groups as they are inherently imbedded in their own (typically white middle-class) cultural, social, political and economic construction of reality.[39]

When urban planning caters to different political interests it not only adopts different forms and methodologies but evidently also leads to different results and, thus, in different types of urban spaces. Some forms of urban planning, such as strategic planning[40] and public-private partnerships,[41] are specifically aimed at accommodating the market, thus, leading to a predominance of the

37 Nicolai Steinø, 'Vision, Plan and Reality.'

38 John Friedman, *Retracking America: A Theory of Transactive Planning* (Garden City, NY: Anchor Press/Doubleday, 1973); John Forester, 'Critical Theory and Planning Practice,' *JAPA* 46 (1980): 275–286; Patsy Healey, 'Planning through Debate: The Communicative Turn in Planning Theory,' *Readings in Planning Theory*, eds. Scott Campbell and Susan S. Fainstein (Malden, MA: Blackwell, 1996), 234–258; Patsy Healey, *Collaborative Planning: Shaping Places in Fragmented Societies* (London: Macmillan Press, 1997).

39 Marsha Ritzdorf, 'Feminist Thoughts on the Theory and Practice of Planning,' *Readings in Planning Theory*, eds. Scott Campbell and Susan S. Fainstein (Malden, MA: Blackwell, 1996), 445.

40 Jerome L. Kaufman and Harvey M. Jacobs, 'A Public Planning Perspective on Strategic Planning,' *Journal of the American Planning Association* 53.1 (1987): 23–33.

41 Gregory D. Squires, 'Partnership and the Pursuit of the Private City,' *Urban Life in Transition*, eds. Mark Gottdeiner, and Chris Pickvance (Thousand Oaks, CA: Sage Publications, 1991), 196–221.

market value over the use value of urban space. As such, these forms of urban planning tend to promote smooth space over friction and, more importantly, to accommodate consumers rather than artists, activists and other disruptive users of urban space.

Regardless of the political aims of urban planning, some uses and users of urban space are intrinsically adverse to planning. As the goal of urban planning, regardless of its political aims, is to organise space to cater to specific uses, planning for disruptive use is a contradiction in terms. Whenever planning aims at doing so it ends up domesticating the disruptive uses and rendering them harmless, i.e. undisruptive, hereby losing their meaning. As an example, the dual aim of facilitating the practice of skating and avoiding its clash with other uses of urban space has led to the creation of skater parks where skaters may practice their skills in an environment that is specifically designed to this end. Hence, skating is tamed and the skater is domesticated. The same is done for graffiti artists who get their own reservations where they are permitted to lawfully perform their art.

In this process, the smoothness of urban spaces is preserved and yet other heterotopias are added to the series of mono-functional spaces which have come to make up contemporary urban space in order to prevent friction and foster smoothness. It is questionable, in other words, whether urban planning should aim at catering for disruptive uses and whether it is at all capable of doing so would it have this aim. On the one hand, the disruptive activities change their nature and are no longer the same. On the other, those performing the disruptive activities would ultimately have to accept domestication. Otherwise they would probably just run off to skate somewhere else.

Ironically, when it comes to disruptive uses, urban planning may be incapable of fostering those Lefebvrian spaces of 'the other' and thus to build an inclusive, tolerant and stimulating urbanity which might otherwise be among its aims. On the contrary, urban planning seems to have to accept that some of the things that it was designed to create may in fact only happen if it stays away. Furthermore, it seems that it will even have to accept uses which may be adverse to planned uses, urban planning's legitimate children, in order to fulfil its purpose.

6 Parkour as Lens

Parkour, also known as the art of movement, is not just the spectacular activity of individuals jumping from roof to roof so prevalent in YouTube clips. It is in fact a serious practice that promotes exceeding individual barriers, both physical and mental and most often in concert.[42] One of the key notions of parkour is the location of its practice. Although originally founded as a military training practice with its ultimate place of practice the jungles of Vietnam, parkour today represents an alternative urban practice along with skating, longboarding or sleeping in public.[43]

Even though specific parkour 'training grounds' now exist, another example of domestication, the real place of parkours' unfolding is in the spaces of the city.[44] The practitioners of parkour are not willing to accept the rules of attire, attendance and attitude dictated by fitness centres indicators of exclusion. As with Lima e Ferreira's example of citizens not being allowed inside government buildings because they were too poor to have 'the right' clothes,[45] contemporary practitioners of parkour take the city as their 'playground.' Their practice re-signs space and exemplifies literally but also symbolically free movement and free expression in public space.

The manner in which the practitioner of parkour – the *traceur* – uses the city is exemplary of ways in which spaces can be re-signed. There is a lack of adherence to the codes of behaviour dictated by specific spaces and specific objects. At every instance there is a possibility of re-signing – a bench becomes the site of a handstand, legs swing over backrest, no sitting down. This notion of re-signing could be akin to Mojanchevska's notion of re-thinking and Topaloğlu & Çebi's re-creating and perhaps a 're-scripting' in the case of Lima e Ferreira – all ways of activating new perspectives and finding new meanings.[46] The movement of the traceur re-defines objects and draws them into constellations with previously unrelated objects. A jump from a curb can lead to a swing through a railing, which in turn can lead to a leap to a tree branch and a shoulder roll through a lawn to be followed by a running along a parapet wall

42 Shelley Smith, 'A Hop, Skip and a Jump: Examining the Experiential Potential in Contemporary Urban Public Space,' *Architecture and Stages of the Experience City*, ed. Hans Kiib (Aalborg: Aalborg University Skriftserie 30, 2009), 347–353 and Shelley Smith, 'Discovering Urban Voids and Vertical Spaces,' *Performative Urban Design*, ed. Hans Kiib (Aalborg: Aalborg University Press, 2010), 146–154.

43 Shelley Smith, 'Hop, Skip and a Jump,' Shelley Smith, 'Discovering Urban Voids and Vertical Spaces' and Ilona E. Gerling, Alexander Pach, and Jan Vitfeld, *The Ultimate Parkour and Freerunning Book* (Aachen: Meyer and Meyer Sports Verlag, 2013).

44 City Surfers, dir. Kasper Astrup Schröder, Copenhagen: KSPR and Streetmovement, 2007. DVD. and Shelley Smith, 'Hop, Skip and a Jump,' Shelley Smith, 'Discovering Urban Voids and Vertical Spaces,' and Unpublished interview: Jeppe Skovgaard, Streetmovement. Interviewed by Shelley Smith. Streetmovement offices, Copenhagen, Feb. 10, 2010.

45 Lima e Ferreira, in this volume.

46 Mojanchevska, Topaloğlu/ Çebi and Lima e Ferreira, in this volume.

to the edge of a planter. Alternate spatial paths are traced and made visible, appealing to rebellion and imagination. With regard to this an additional link can be made to another chapter in this volume that of Anthi Kosma's 'Spacing by Trace: Four Events of an out of Place Experience.' Kosma relates her 'Four Events' as a kind of 'cartography of the different aspects of the art of drawing'[47] and this parallels the practice of the traceur in the drawing of these alternate urban spatial paths. Kosma notes that although the scales at which the practice of drawing and the practice of the traceur of parkour take place are very different, the practices themselves have similarities in the ways that bodies can reconceptualise space through sensorial aspects.

Implicit in the activity of the traceur is the tactility of space. Space becomes at once both background – the spaces traversed – and foreground – the materiality of stone, grass, concrete coming into contact with the skin of fingers and the rubber of soles. In this practice, the role of 'form' is diminished, fading into the background space of the public and the foreground space of an exchange.

This diminished role of form is in keeping with what urbanist Albert Pope describes as the overwhelming of the architectural object in contemporary urbanity.[48] In this expanse of space, the spatial characteristics of contemporary urbanity challenge the traditional notions of public space that propounded fixity and a focus on the middle ground of formal frameworks for contact: 'Only by abandoning the primacy of built form is it possible to reposition form so that it may effectively respond to a city dominated by space.'[49]

Parkour makes manifest through its practice this condition of contemporary urbanity, one in which space (background) rather than form (middle ground) is the main factor and introduces in an interplay with space, materiality (foreground). In this sense, parkour perhaps provides a clue to ways in which the background of space can be regarded and experienced in addition to how alternative practices can contribute to the understanding of space in a contemporary context and to the creation of rich and diverse places.

In this regard, the use of the terms space and place are of particular interest here. *In Space and Place: The Perspective of Experience*, Tuan makes the case that space and place are co-dependent, arguing that the ability to move from one place to another defines space while the existence of place requires space. This he put succinctly as '[p]lace is security, space is freedom.'[50] Tuan's distinction can be taken into the discussion of contemporary public space in the

47 Anthi Kosma, this volume.

48 Albert Pope, *Ladders* (Princeton: Princeton University Press, 1997), 3.

49 Ibid., 7.

50 Tuan, *Space and Place*, 3.

context of external political and economic agendas being intent on creating public space as place in a particular 'secure' and foreseeable manner, while the essence of public space may in fact lie in more moveable, dynamic and flexible spatially based models. As such, it offers more possibility for experience. A point also made by Mojanchevska and Lima e Ferreira.

Further to this, the practice of parkour in its choice of the city as the place of practice exhibits a form of ownership. Where urban public spaces can encourage alienation rather than inclusivity and distance from the spaces of the city and its inhabitants, the practitioners of parkour claim the city as theirs, disregarding predetermined functions, codes of behaviour and in general 'what people think.' The city is available to them 24-7. By providing a spectacle of space and movement they contribute to the space of the public and expand the definition of what, how used and by whom.

7 Conclusion

In the post-modern economy, the nature of public space has shifted from an orientation towards use-value and habitation towards exchange value and tourism. In this process, an increasing alienation towards public space is occurring, discouraging some and encouraging no one to claim space.

Architects and urban and landscape designers are not innocent in this process. Whether commissioned by the public or the private sector, they have their focus on the middle ground. This is also the focus of the tourist view, one intent on cleaning up, aestheticising and preparing urban space as a stage for leisure and consumption.

The practice of parkour in urban space focuses on the use-value and appropriation of urban space. In doing so it provides an example of urban ownership a situation in which the city is claimed and inhabited as a space for personal attachment and unfolding and ultimately education. In addition, parkour focuses on both the foreground and the background of urban space making manifest spatial connections and materiality and 'making space' for the possibility of social connectivity and exchange.

While these aspects are not the focus of formal considerations in the middle ground, they are the foci that to a great degree enable public space design to fulfil a function that in contemporary urbanity makes social education through exchange possible. This kind of public space provides a framework for all manner of possible actions and interpretations, uses and functions. It encourages the 'inhabitation' of space by not being fixed and inflexible but by being dynamic and pliant. In order for architecture and design to align with use-value and appropriation of urban space, it will also have to change its focus.

Bibliography

Davidoff, Paul. 'Advocacy and Pluralism in Planning.' *A Reader in Planning Theory*, edited by Andreas Faludi, 277–296. Oxford: Pergamon Press, 1973.

Dovey, Kim. *Framing Places: Mediating Power in Built Form*. London: Routledge, 1999.

Forester, John. 'Critical Theory and Planning Practice.' *JAPA* 46 (1980): 275–286.

Friedman, John. *Retracking America: A Theory of Transactive Planning*. Garden City, NY: Anchor Press/Doubleday, 1973.

Gerling, Ilona E., Alexander Pach, and Jan Vitfeld. *The Ultimate Parkour and Freerunning Book: Discover Your Possibilities*. Aachen: Meyer and Meyer Sports Verlag, 2013.

Hajers, Maarten and Arnold Reijndorf. *In Search of New Public Domain*. Rotterdam: NAi Publishers, Rotterdam, 2001.

Healey, Patsy. 'Planning through Debate: The Communicative Turn in Planning Theory.' *Readings in Planning Theory*, edited by Scott Campbell and Susan S. Fainstein, 234–258. Malden, MA: Blackwell, 1996.

Healey, Patsy. *Collaborative Planning: Shaping Places in Fragmented Societies*. London: Macmillan Press, 1997.

Kaufman, Jerome L. and Harvey M. Jacobs. 'A Public Planning Perspective on Strategic Planning.' *Journal of the American Planning Association* 53.1 (1987): 23–33.

Leach, Neil. *The Anaesthetics of Architecture*. Cambridge, MA: MIT Press, 1999.

Lefebvre, Henri. 'Right to the City.' *Writings on Cities*. Oxford: Blackwell, 1996.

Lofland, Lyn. *The Public Realm: Exploring the City's Quintessential Social Territory*. New York: Aldine de Gruyter, 1998.

Pope, Albert. *Ladders*. Princeton: Princeton Architectural Press, 1997.

Ritzdorf, Marsha. 'Feminist Thoughts on the Theory and Practice of Planning.' *Readings in Planning Theory*, edited by Scott Campbell and Susan S. Fainstein, 445–450. Malden, MA: Blackwell, 1996.

Schröder, Kasper Astrup and Street Movement. *City Surfers*. Directed by Kasper Astrup Schröder. Copenhagen: KSPR and Streetmovement, 2007. DVD.

Smith, Shelley. 'A Hop, Skip and a Jump: Examining the Experiential Potential in Contemporary Urban Public Space.' *Architecture and Stages of the Experience City*, edited by Hans Kiib, 347–353. Aalborg: Aalborg University Skriftserie 30, 2009.

Smith, Shelley. 'Discovering Urban Voids and Vertical Spaces.' *Performative Urban Design*, edited by Hans Kiib, 146–154. Aalborg: Aalborg University Press, 2010.

Smith, Shelley. Streetmovement. Unpublished interview: Jeppe Skovgaard, Streetmovement. Interviewed by Shelley Smith. Streetmovement offices, Copenhagen, Feb. 10, 2010.

Squires, Gregory D. 'Partnership and the Pursuit of the Private City.' *Urban Life in Transition,* edited by Mark Gottdeiner and Chris Pickvance, 196–221. Thousand Oaks, CA: Sage Publications, 1991.

Steiner, Ditmar. 'Die Stadt zwischen der Sehnsucht nach Harmonie und den Gesetzen der Dienstleistungsgesellschaft.' *Risiko Stadt? Perspektiven der Urbanitet,* edited by Ullrich Schwarz, 216–221. Hamburg: Junius, 1994.

Steinø, Nicolai. 'Vision, Plan and Reality: Urban Design between Conceptualization and Realization.' PhD dissertation, Aarhus School of Architecture, 2003.

Tuan, Yi-Fu. *Space and Place: The Perspective of Experience.* Minneapolis: University of Minnesota Press, 1977.

Zukin, Sharon. 'Postmoderne urbane landskap: A kartlegge kultur og makt.' *Pa sporet av Byen,* edited by Jonny Aspen and John Pløger, 159–190. Oslo: Spartacus, 1997.

Zukin, Sharon. 'Whose Culture? Whose City?' *The City Reader,* edited by Richard T. Le Gates and Frederic Stout, 131–142. 2nd ed. London: Routledge, 2000.

CHAPTER 6

The Urban Challenge of Participation through Theatre in Rio

Ana Carolina Lima e Ferreira

Abstract

Rio is a contrast of beautiful landscapes and seaside dreams and violence, danger and lack of infrastructure. It is a city where people suffer from inequalities. The Theatre of the Oppressed created by Augusto Boal (1931–2009) is an initiative crossing physical and cultural boundaries, giving voice to people and allowing urban transformation. This method was developed when he was exiled during the Brazilian Dictatorship. In the 1980s Boal returned to Brazil and in the 1990s he was elected a city councillor in Rio de Janeiro. His 'Legislative Theatre' created a platform where many suggestions were received from the people to formulate new and coherent projects of law. With public participation, ideas were collected and transformed into Boalian theatre interventions in public spaces. Through this approach, Boal implemented fifteen city laws and two national laws. Nowadays the people of Rio are still suffering due to urban transformation. The theatre group, however, still continues their actions by democratising cultural products providing possibilities to activate and strengthen the citizenship. They still believe in the importance of implementing projects, activating participation and turning the oppressed into protagonists – creating a dialogue through aesthetic media. Art reframes urban intervention into public art pieces; theatrical participation into empowering performance; and the work of community leaders into a process for urban change.

1 Introduction

Public Art in the form of the Theatre of the Oppressed connects audiences in public spaces also unifying the people and urban space as whole. These theatrical approaches also allow people to intervene on the scene and raise questions of oppression. With 'Legislative Theatre' the citizens also have the chance to influence public decisions. Boal's Theatre of the Oppressed reveals oppression through theatre play and at the same time locates ways to overcome the problem of oppression. The actors are people, who themselves feel oppressed. They

© KONINKLIJKE BRILL NV, LEIDEN, 2018 | DOI 10.1163/9789004365520_008 could be professionals or non-actors – common people. Boal's method gives voice to those who claim for urban transformation against oppression. Boal wrote many books systematising his method. His masterpiece, *The Theatre of the Oppressed*, was published in Argentina in 1973 while he was in exile escaping from the Brazilian Military Dictatorship (1964–1985). While in exile Boal was still active, practicing and sharing his experiences with theatre. He spent most of his exile in Paris teaching Theatre of the Oppressed at the Paris Sorbonne University and facilitated many workshops around Europe, spreading his method worldwide. He affirmed that art not only presents a way to see the world but can also transform it. Boal worked with the concept of the estrangement effect, *Verfremdungseffekt,* which influenced his focus to create dialogue about oppression.[1]

The plays produced and performed by community members can be seen as a form of urban intervention, as public art pieces. The theatre's participation of non-actors creates a community and a process for urban changes. As such, this chapter resonates with some other successful cases of urban change described by other authors in this volume: Katerina Mojanchevska,[2] İklim Topaloğlu and Pelin Dursun Çebi[3] and by Shelley Smith and Nikolai Steinø.[4] The methodology of my research is qualitative: observation of the theatre plays; participation of Theatre of the Oppressed workshops; analysis of case studies in various cities such as Rio de Janeiro; and interviews with participants and experts from urban spaces as well as actors and directors. Based on interviews with Olivar Bendelak, one of the coordinators of the Centre of Theatre of the Oppressed in Rio (CTO Rio), this chapter analyses projects in Rio considering the material, the artistic and urban sociological context and the contemporary problems related to them.

2 The Theatre of the Oppressed in Praxis

Augusto Boal developed a method of popular education for emancipation specifically to give voice to the oppressed, encouraging them to identify and claim

1 Augusto Boal, *Teatro do Oprimido e outras poéticas artísticas* (Civilização Brasileira. Rio de Janeiro, 2005) 31.

2 Katerina Mojanchevska, 'Self-Organised Citizens' Groups and Urban Space: Challenges in Planning Paradigm,' in this volume.

3 Iklim Topaloğlu and Pelin Dursun Çebi, 'New Interventions in Public Space: Spatial Readings through Art and Architectural Installations,' in this volume.

4 Shelley Smith and Nicolai Steinø, 'Public Space Design between Alienation and Appropriation: The Case of Parkour,' in this volume.

their objectives. The theatre is used as a way to examine cultural differences and urban inequalities. Theatre of the Oppressed was influenced by academic pedagogue Paulo Freire, author of the book *Pedagogia do Oprimido*, which analysed oppression based on Brazilian inequalities. Many others inspired Boal, such as Bertolt Brecht and Konstantin Stanislavsky for their theatrical work with actors. Boal blended many theatrical concepts to create a method to empower the oppressed though theatre. Fundamentally, Theatre of the Oppressed provided a medium for political engagement and political struggle in the service of the oppressed. During his exile, Boal was active in sharing his experiences on theatre and spreading his method. The method is now used worldwide with practitioners adding new techniques and adapting to respond to local necessities.

CTO Rio is the reference point of Theatre of the Oppressed in Brazil and is also becoming so in Latin America. The main goal of CTO Rio is to further develop the Theatre of the Oppressed in Brazil and Latin America. Boal was its art director from 1993 to his death 2009. It is a centre for research, experience and diffusion of his method, as well as the analysis, systematisation of exercises, games and theatre techniques. It is also a location for seminars and socio cultural projects, theatre plays and art products based on the Aesthetic of the Oppressed, which is the principle to democratise art developed by Boal. The goal of those actions is to generate cultural products from the communities as possibilities of intellectual development of the Media and to activate and strength the citizenship. CTO Rio implements projects together with many communities formed by oppressed groups mostly by inhabitants of the poor areas, and encourages active participation to turn the oppressed into protagonists with the aim of creating dialogue through aesthetic media. The Theatre of Oppressed in Rio facilitates daily action for societal debate. Boal's book *Aesthetics of the Oppressed* focuses on the ideas and experiences with communities developed at this location in Rio. The principles of his aesthetics are to give voice to the oppressed by using all arts as a tool for transformation. Therefore, it allows people to become aware of their oppression and do something to change it. The CTO Rio was a meeting point and an exhibition space for such projects which developed from Boal's principles in the *Aesthetics of the Oppressed*. Working with diverse partners, institutions and schools, Theatre of the Oppressed spread throughout Brazil with other theatre projects such as de *Ponto a Ponto*. All of these projects followed the principles of the *Aesthetics of the Oppressed* to develop and grow. It is another possibility to share the knowledge of the Theatre of the Oppressed with more people in other regions of Brazil, allowing more people to participate and propose solutions concerning Brazilian problems. These actions come from their own views and not top down. By doing so, the citizens throughout Brazil have the chance to re-take and define the urban spaces based on their necessities as is also discussed by Smith and Steinø.

3 Theatre Director as a Member of the City Council of Rio de Janeiro

This centre CTO Rio was a key point during Boal's mandate as a city councillor of Rio de Janeiro. It was where his staff worked to set the theatre plays and discuss the issues. In the most popular form of the Theatre of the Oppressed, 'Forum Theatre,' a situation of oppression is presented in scenes coming from a concrete problem, out of 'real' scenes of oppression suffered by the participants. The audience has the chance during the scene to replace the actors to frame their own questions for discussions through theatrical activities and after to have a collective debate and evaluation. After being elected as city councillor, Boal developed a new way to practice the Theatre of the Oppressed in direct relation with politics. Called 'Legislative Theatre,' it allowed participants to create proposals for laws, to be approved by the members of the City Council after a scene of 'Forum Theatre.' After the play prepared by his theatre group there was a debate and people from the audience could replace the actors in a situation of oppression that mimicked a concrete social problem. After the theatre plays the public had the chance to give suggestions for new laws out of these scenes of oppression.

Boal's goal with 'Legislative Theatre' was to formulate new and coherent projects of law. He and his theatre group used this method to implemented fifteen city laws and two national laws. As a response to the forced political apathy of the 1990s in Brazil caused by the military dictatorship, Boal's 'Legislative Theatre' group used public participation for creating legislative power. According to Bendelak, it was a unique experience in the world and it produced concrete effects.[5] They worked a lot with oppressed communities and formed around 60 groups of 'Forum Theatre' and 33 remained quite stable for a long time.

In his campaign to be city councillor in the elections in 1992, Boal extensively used Theatre of the Oppressed as a part of the election campaign around Rio – a theatrical campaign, a campaign through theatre. His goal was to use Theatre of the Oppressed to create legislative power. And so began Boal's 'Legislative

5 Olivar Bendelak, Interview with the Author that is coordinator of Legislative Theatre at the Centre of Theatre of the Oppressed Rio de Janeiro, on 17th of January 2012.

Theatre.' The campaign was successful and after many theatre shows Boal won the election. For him the legislator is not supposed to be the person who makes the law but the person through whom the law is made. Boal worked intensively to give voice to the oppressed to transgress, to break the conventions, to enter into the mirror of a theatrical fiction, rehearse forms of struggle and then return to reality with images of their desires. He considered that this discontent was the genesis of the 'Legislative Theatre' in which the citizen makes the law through the legislator.[6]

One area of discontent in Rio is the multitude of problems surrounding public health. The public hospital and medical centres are overcrowded and there are not enough trained personnel. For the rich, there are the private hospitals and clinics with better quality service which is extremely expensive. For the poor, who cannot afford private health insurance, the only option is the public health system with its overcrowded public hospitals. Boal mandated one specific health problem with aged people. The group needed a specialist medical doctor for their health problems: a geriatrician. However there were no geriatric doctors in the public hospitals in Rio. Sometimes aged patients were even assisted by paediatricians. Boal's theatre group used theatrical intervention to rectify this situation. When the law was approved by the city councillors, first the mayor vetoed it. The mayor at that time, César Maia, argued that there were geriatric doctors in the city hospitals because the State gives money to the city in order to hire the geriatrician. However this showed that the resource management, a mayoral responsibility, was not efficient and which was further impeded by the fact there was no law obliging the city to specifically hire a geriatrician. As a city council member Boal took a group of aged people to the biggest city hospital in Rio, the Souza Aguiar hospital. There they did a theatre intervention pretending they wanted medical assistance and wanted to see the geriatric doctor. Then the receptionist said that the hospital did not have any geriatric doctors and could find only a general doctor. Then Boal, as city councillor and a representative of the aged group, talked to the hospital director and the press about the lack of geriatricians. The matter of law was back on the agenda of the City Council and the veto was dismissed. Through Boal's theatrical intervention the law was approved and now all city hospitals in Rio are obliged to have a geriatric doctor.

With 'Legislative Theatre' Augusto Boal had the idea to give transparency to what happens inside the City Council Hall. Unfortunately, due to the violent years of dictatorship, the Brazilian population is at times somewhat apathetic

6 Augusto Boal, *Legislative Theatre* (Routledge, Oxon, 2006) 9.

in its practice to follow politics. People do not go the City Council or to the State Parliament to see a voting session as it happens in other countries. So Boal did an event called *Câmara na Praça*, (City Council in the Square) in front of the City Council Hall in Cinelândia in the centre of Rio. They built a stage there and Boal started talking about the projects, which were under discussion for vote each week. Then the theatre group presented a scene of 'Forum Theatre' using themes from the community. After the scene of 'Forum Theatre' the citizens discussed it. Sometimes, when there was a very important issue to be voted upon in the City Council Hall, another *Câmara na Praça* was organised and would finish only a few minutes before the voting would start. This encouraged the audience to enter the building to watch the session creating the opportunity to see the debates after the votes. Prior to this theatrical intervention the audience did not know they were allowed to enter and see the political debate. Another obstacle to entering the Council was that many times people were dressed too casually, as people normally do, because Rio has an average temperature range of 28–30° c. Such a simple rule as a dress code was enough to exclude the citizens to participate politically. It is what Smith and Steinø describe as marginalisation of users. Thinking on it Boal's group brought to the event theatre costumes in different sizes: elegant shirts, jackets, ties and elastic pants and skirts that would fit almost anybody. The public borrowed the clothes for a while and so had the right dress code and could enter the building.

4 Who are the Oppressed in Rio de Janeiro?

Olivar Bendelak affirms that the poor people are the oppressed in Rio, mostly the inhabitants of the poor areas on the city's peripheries or *favelas* who suffer from the lack of basic infra-structure as sewage systems, potable water and waste management. The transportation system is not good for the ones who live in the distant poor areas and go to work in the richer areas in the southern parts of Rio. Moreover, the population in the *favelas* as a whole is constantly in danger, hostage either to the police who criminalise the poor or to the drug dealers who control the area with guns. The poor population also suffer economically with low salaries and other kinds of economic oppression caused by unemployment or underemployment, lack of education or professional training, bad living conditions, other the difficulties with the location and violence. The *favelas* in the north areas are mostly precarious homes, barracks made of wood and some made of cardboard which exposes people to danger and health problems. Nevertheless, Bendelak is full of hope and believes that the situation has been improved with the current presidential government of Partido dos Trabalhadores, the Workers Party.[7]

5 The Use of Public Space

Since Boal passed in 2009, CTO Rio has had a series of difficulties to keep the work of the Theatre of the Oppressed going. Bendelak explains that in the time of Boal's mandate the group CTO Rio could go everywhere, every square they used a tent and performed theatre. Recently the municipality of Rio, using the excuse of coordinating the public spaces, requires the group to gain authorisation from the municipality to use the spaces of the administration region, the central area, even if it is a square with an amphitheatre. Furthermore, there is renovation being done on the pavements in the area Lapa, where CTO Rio is located, of which the group was not informed. Bendelak states that the authoritarian attitude impedes the group's function and space. The group also used to present plays in the area of Largo da Carioca in front of a subway station but now the group is only allowed to present in a corner of the station with fences surrounding it. This was an ideal space for the group to have an impact as thousands of people circulate there from Monday to Friday. But now this public space is restricted. The issues of cleaning up the urban space and marginalising some users, mostly the poor, are also discussed in the chapter by Smith and Steinø.

Bendelak believes it is necessary for the CTO group to facilitate accessibility to cultural activities, music, dance and circus etc. Rio has great spaces which should be used for culture.[8] While the group can ask for permission, this impacts on the necessary spontaneity of the theatrical method. In Brazilian culture the spontaneous use of space concerns the whole Brazilian way of life. The climate is favourable and public spaces are supposed to be places free of charge for everyone as is the case with the beaches or parks etc. Moreover, if the group comes up with a spontaneous demonstration to claim for an urgent cause, it is almost impossible to get permission at the time because of the slowness of Brazilian bureaucracy.[9] It takes, in fact, at least five days with all the bureaucracy involved. The theatre even performed without permission and was forced to leave a square by the police. The group would love to always be in public spaces in Rio but for that they need support from the municipality,

7 Bendelak, Personal Interview.
8 Ibid.
9 Ibid.

which does not exist. According to the Law School Fundação Getúlio Vargas-Rio, the general rule is that public spaces can be used to develop the purpose for which they are intended: the school for the provision of education services; the hospital for health care of the population and so on. Bendelak affirms that the idea of the Public Square for the stage of manifestations and demonstrations is slowly disappearing as the government is promoting big events such as the Olympic Games and the Football World Cup. Right now the local administration is building huge stages for concerts and paying enormous amounts of money to certain artists to stage concerts even in the peripheral poor areas of Rio as part of election campaign promises. Meanwhile the Theatre of the Oppressed and the CTO Rio is disregarded, maybe because it promotes the capacity to use art to express, to reflect and to fight oppression.[10]

As in Self-Organised Citizens' Groups in the chapter by Mojanchevska, the aim of the Theatre of the Oppressed is very well framed. The oppressed get the chance to expose through theatre, their oppression in their own way. It is considered very successful and is practiced in many countries, changing through the years, adapting to each regional context – the aim is still the same, but how and by whom it is coordinated is done differently. There is still no top down control, meaning no bureaucratic regulation. Nobody needs authorisation or previous formation in theatre to play or to take part in this theatre.

6 The Achievements

Bendelak affirms that the CTO Rio made many legislative gains concerning the rights of disabled people, aged people and many other minority groups with 'Legislative Theatre.' Another key achievement that is not generally measured in economic terms is how the theatrical movement gave voice to people who were oppressed. These transformations are achieved on the level of citizenship. For Bendelak the key achievements are the concrete actions.[11] The main goal of the Theatre of the Oppressed in his opinion is not just to perform, show the scene, but to have action.[12]

As an example, in Rio there are between 9,000 and 10,000 maids but only around twenty percent have a stable position with guaranteed labour rights. One branch group of Theatre of Oppressed, *Marias do Brasil*, was formed in 1998 by ten domestic servant women to fight this economic oppression. The

10 Ibid.

11 Ibid.

12 Ibid.

popular group accumulated extensive theatre experience in operation, having two spectacles of 'Forum Theatre' which participated in national festivals of Theatre of the Oppressed. *Marias do Brasil* also had shows included as part of the World Social Forum. At the Festival of Associative Theatre Federation in Rio de Janeiro, FETAERJ, Maria Vilma, a member of the group, received an Honorable Award. With projects concerning Human Rights, such as the plays *Encenando Direitos Humanos* and *Maria Luta por Lei Justa,* the theatre expands its reach and the audience discusses and proposes laws concerning working rights.[13] Every Brazilian middle and upper class home has a domestic servant woman, mostly underpaid, who does much more than a housekeeper: from cooking, cleaning, laundering, ironing, babysitting, gardening, fixing things and undertaking all other household duties. As Provisional Law, Lei Federal [Federal Law], 10.208 de 23/03/2001, determines the optional payment of the golden handshake for domestic servants. *Marias do Brasil* and the Unions of Domestic Servants of the city of Rio de Janeiro and Nova Iguaçu claim that this measure has to become mandatory. For this they started a movement to collect signatures and petition Congress. The theatre group invited all maids to join the cause, perform in a scene and support their petition.[14] Finally, *Marias do Brasil* started to valorise its group of domestic servants who gained self-esteem since they started Theatre of the Oppressed. The project is ongoing and continues to fight against the injustices domestic servants suffer by advocating. Along with CTO Rio they even presented a theatre play in National Congress in Brasília, the Brazilian Parliament, to claim for better rights for their underprivileged working class. The Theatre of the Oppressed offers justice but it is necessary to first improve self-esteem with real inclusion through concrete theatrical actions.

Such situations like the domestic servant's movement is very similar as well to those points discussed in Mojanchevska's chapter on the Self-Organised Citizens' Groups in this volume.

7 How to Democratise the Theatre of the Oppressed in Rio?

The goal of the Theatre of the Oppressed is to democratise and reframe the media and cultural products as possibilities of intellectual development in order to activate and strengthen the citizenship. To democratise arts in Rio is a

13 Centro do Teatro do Oprimido Rio de Janeiro CTO Rio. Viewed on 30 April 2015. http://ctorio.org.br/novosite/o-que-realizamos/grupos-populares/.

14 Ibid.

challenge and the main reason is the social inequality. In order to implement projects, CTO Rio needs funding as there are not many resources available for art and culture. Moreover, the group has a philosophy of work based on ethics and solidarity which prevents it from taking sponsorship from sectors or companies that exploit or oppress people. Fortunately they currently receive funding from government money and can work with total freedom. The financial problem is a serious one because they always have to search for sources, sometimes only small projects can be funded and it makes their work very hard. However, the mission of the Theatre of the Oppressed is still needed, in fact, as some problems go away others come in.

David Harvey affirms that the question of what kind of city we want cannot be divorced from what kind of social ties, relationship to nature, lifestyle, technologies and aesthetic values we desire.[15] He says that the right to the city is far more than the individual liberty to access urban resources – it is a right to change ourselves by changing the city. For Harvey, one step towards unifying these struggles is to adopt the right to the city as both working slogan and political ideal precisely because it focuses on the question of who commands the necessary connection between urbanisation and surplus production of use.[16] For him the democratisation of that right and the construction of a broad social movement to enforce its will are imperative if the dispossessed are to take back the control, which they have for so long been denied, and if they are to institute new modes of urbanisation. For Harvey, Henri Lefebvre was right to insist that the revolution has to be urban in the broader sense of the term or nothing at all.[17]

8 Performance, Space and Potential

The general picture of the Theatre of the Oppressed is to fight against the social inequalities. In Rio, 'Legislative Theatre' was created to allow people to participate with suggestions of law during the time Boal was a city councillor. Nowadays the mission and actions of Theatre of the Oppressed are extremely necessary to fight for social justice. It can be seen that the method in the praxis has a lot of challenges but amazing results have been achieved on the activation of citizenship, empowerment and giving voice to the oppressed. Like Lefebvre asserts, the spaces of representation are lived, it has a core or an affective

15 David Harvey, 'The Right to the City' *New Left Review* 53 (2008): 40.

16 Ibid.

17 Ibid.

centre, it contains the environments of passion and action, those situations experienced thus immediately implies time. So the space of representation, a theatrical space, can receive various qualifications: directional, situational, relational, because it is essentially qualitative, fluid and dynamic.[18]

The dialogue with the plurality is the most remarkable public art characteristic, which happens normally through a collective project and interdisciplinary focus on urban space. This also resonates in the discussions presented in the chapter by Topaloğlu and Dursun Çebi and also in the chapter by Smith and Steinø. Art in public spaces is hard to define. It is very heterogeneous, belonging to different social layers with diverse cultural education. Thus, one of the main public art goals is to establish dialogue with diversity, which is very challenging for the artist working in an urban environment. Smith and Steinø assert that the traditional public space encourages plurality. Lefebvre affirms that dominant spaces are the centres of richness and power, which strives to shape the dominated spaces – those of the peripheries. It is reduced to an action often violent to obstacles and the resistance.[19] Lefebvre suggests also that the reality called social is double, multiple and plural.[20] Smith and Steinø highlight the fact that, ironically, when it comes to disruptive uses, urban planning may be incapable of fostering those Lefebvrian spaces of 'the other' and thus to build an inclusive, tolerant and stimulating urbanity which might otherwise be among its aims.[21] When measured, it is not itself anymore and it is not in a way of materiality. It contains, it implies a concrete abstraction (money, commodities, exchanges of goods), thus the 'pure' forms as those of exchange, of language, of signs, of equivalents, reciprocities, contracts, etc.[22] Social spaces are *produced*, from a 'raw material,' nature. As in the case of Theatre of the Oppressed, social spaces are produced by an activity, which implies an economy, a technique, but goes far beyond – producing politics and strategies.[23] Social spaces result in a process of multiple aspects and movements: significant and non-significant, perceived, experienced, practical and theoretical.[24] The space, finally with its double determination, is at the same time fictional, produced and productive, social-material, immediate and mediation

18 Henri Lefebvre, *La production de l'espace* (Paris: Anthropos, 2000), 52.

19 Ibid., 61.

20 Ibid., 41.

21 Smith and Steinø, in this volume.

22 Lefebvre, *La production de l'espace*, 98.

23 Ibid., 100.

24 Ibid., 130–131.

(milieu and transition) connection and separation.[25] Lefebvre's theory of double determination helps explain the theatrical space where the play with the double determinants of fictive and real, the interference of look and mirages has a reencounter with the actor, the public and the 'character,' the text and the author. This play makes the bodies go to a 'real' space immediately lived (the room, the scene) to a 'perceived' space and a third space, which is not the scenic space or a classic theatrical space. The theatrical space implies a representation of the space, the scenic space, which corresponds to a conception of space. The space of representation is mediated but lived, enveloping a work in the moment of the play according to Lefebvre.[26] For him to liberate the basics and fundaments on which they can build in a genetic process, spaces of various societies are only the beginning of an exploration of this 'reality' apparently translucent. In the context of the new urban space under Lefebvre's concept of the production of space there is a difference between the kind of users and their antagonist interest in that space. Lefebvre's assertions, then, go some way to explain the social spaces of Rio and the potential of socially engaging work like that done by the Theatre of the Oppressed.

9 Conclusion

> The citizen who in the everyday life has been reduced to a partial function (means-ends activities) can be discovered in art as 'human being.'[27]

The main focus of this chapter is the social critic and the use of public space. It is possible for the citizen to be more visible and give voice to the community through active participation. Art can be socially engaged using the artist's creativity as well but it can make the citizen participant role more active and stronger. The Centre of the Theatre of the Oppressed Rio therefore implements projects and active participation to turn the oppressed into protagonists, creating dialogue and action. Boal's use of 'Legislative Theatre' is an example of this in the urban space resulting in the approval of thirteen city laws. Furthermore, as Bendelak affirms, the Centre of the Theatre of the Oppressed in Rio had

25 Ibid., 217.

26 Ibid., 216.

27 Peter Bürger, 'The Negation of the Autonomy of Art' *Participation*, ed. Claire Bishop (London: Whitechapel Gallery, and Cambridge: The MIT Press Cambridge, 2006), 47.

achievements on the level of citizenship creating motivation among the oppressed to participate.[28]

Therefore, the transformations created by the Theatre of the Oppressed on the level of citizenship and the motivation to participate, were its fundamental achievements. The theatre group creates spatio-corporeal interventions to give voice to those participants. The techniques created by Boal has many applications but is mostly used for social and political struggle. Rio cto never strays from its fundamental purpose to allow oppressed people to fight. Boal believed that citizens develop their taste for political discussion (democracy) through their desire to develop their taste for political abilities (popular art).[29] He affirmed that the shows contained moments of dialogue, exchange, learning, teaching and pleasure.[30] Umberto Eco's concept of open work suggests a work of art is a complete and closed form in its uniqueness as a balanced organic whole, while at the same time constituting an open product on account of its susceptibility to countless different interpretations which do not impinge on its unadulterable specificity. Therefore, for him every reception of a work of art is both an *interpretation* and performance of it because in every reception the work takes on the fresh perspective for itself. As Eco asserts, the Theatre of the Oppressed is an open work which allows for multiple interpretations. Theatre is grounded in presence and interpretation but Boal's theatrical method goes further to democratise arts, the means of the production and helps fight against inequality.

Bibliography

Baumann, Till. *Von Politisierung des Theatres zur Theatralisierung der Politik. Theatre der Unterdrückten im Rio de Janeiro der 90er Jahre*. Stuttgart: Verlag, 2006.

Bishop, Claire, ed. *Participation*. London: Whitechapel Gallery and Cambridge, MA: MIT Press, 2006.

Boal, Augusto. *Teatro do Oprimido e outras poéticas artísticas*. Rio de Janeiro: Civilização Brasileira, 2005.

Boal, Augusto. *Estética do Oprimido*. Rio de Janeiro: Garamond, 2009.

Boal, Augusto. *Hamlet e o filho do padeiro: memórias imaginadas*. Rio de Janeiro: Editora Record, 2000.

Boal, Augusto. *Legislative Theatre*. Oxford: Routledge, 2006.

28 Bendelak, Interview.

29 Boal, *Legislative Theatre*, 9.

30 Ibid., 9.

Freire, Paulo. *Pedagogia do Oprimido*. Paz e Terra, São Paulo, 2005.

Harvey, David. 'The Right to the City.' *New Left Review* 53 (2008): 40–44.

Heritage, Paul. *The Courage to Be Happy: Legislative Theatre and the 7th International Festival of the Theatre of the Oppressed.* TDR 38.3, The MIT Press. Autumn, 1994. Viewed 15 May 2015. http://www.jstor.org/stable/1146376.

Leavy, Patricia. *Method Meets Art: Arts-Based Research Practice.* NY: The Guilford Press, 2009.

Lefebvre, Henri. *La production de l'espace.* Paris: Anthropos, 2000.

Silva, Fernando P. *Arte Pública: Diálogo com as Comunidades*. C/Arte, Belo Horizonte, 2005.

CHAPTER 7

New Interventions in Public Space: Spatial Readings through Art and Architectural Installations

İklim Topaloğlu and Pelin Dursun Çebi

Abstract

Public space can create social memory and enliven cities through the use of constructed spatial elements, streets and squares, as well as through the spatial events and interventions that take place in these particular spaces. With the help of these elements, urban space becomes its own open-air exhibition. People from all walks of life can create an ongoing communicative relationship with the space and with the city. Today, a new design culture is stimulating production, action and interaction. Multiple and elusive meanings, a pluralistic environment and the participation of the audience in design are emphasised. In the discipline of art and architecture we are moving from static, definite and finished icons to more open, changing and reflective spatial entities. New forms of art and new architectural installations in urban spaces provide an experimental medium for citizens to communicate with their surroundings. They can re-structure their spatial perceptions and reconceptualise the spatial structure of the city. By disturbing usual everyday practices, rhythms and choreographies, these new art forms propose alternative living scenarios in public space. It is these creative spatial interventions that enliven both the body and the city. Resonating with the work of Shelley Smith, Nicolai Steinø, Ana Carolina Lima e Ferreira and Katerina Mojanchevska in this volume, the aim of this chapter is to discuss the contributions of participant and audience-oriented art and architectural installations to urban public space. Selected installations are examined in terms of their spatial potential, their ways of transforming public space and the tactics that are used for the redefinition and recreation process. It is believed that decoding the design features of these sophisticated installations will enhance our spatial awareness and help both designers and architects to create more such innovative concepts and, in the process, encourage the re-definition of public space.

1 Introduction: Bodily Experience in Public Space

Urban public spaces are designed with the interest, welfare and access of the public in mind. They exist to serve the common good. At the same time, public

 | DOI 10.1163/9789004365520_009

spaces are the ultimate space for encounter and interaction, a place for the experience of the body to take form and shape. In this respect, we can talk about the conception of urban life that has evolved through the contribution of the body. The body is widely accepted as a tool for creating our lived experience in the city. It is the centre of agency, a 'location for speaking and acting on the world.'[1] However, in Smith's and Steinø's work[2] it is mentioned that in consumption spaces, body movement and expression are limited by setting rules about when, how and by whom public space can be used. These compulsory public behavioural codes cause conflict and friction.

According to Merleau-Ponty, the notion of spatiality is presupposed by the existence of our body. He considered the human body as our decisive connection to the world. We interact and conceptualise our surroundings with the help of our body. Here, the body is not a static entity, it is an 'acting and performing body' that perceives its surroundings and gains awareness in relation to space, time and movement.[3]

Alva Noe emphasises the idea that perceiving is a way of acting: 'perception is not something that happens to us, it is something we do.'[4] Michel De Certeau, in line with the ideas of Noe, describes space as a 'practiced place.'[5] According to him, geometric areas are transformed into spaces by the practices of their inhabitants. Spatial practices of the body determine the structure of the conditions of daily life and provide a means of communicating with our environment. De Certeau describes the practices we use in everyday life as part of a theoretical model similar to that of linguistics. Like speakers, users manipulate and transform their language and their everyday practices to adapt to their circumstances.[6] By using their own bodies, people create their own practices, meanings and relations. De Certeau calls this 'a way of thinking

---

1 Setha M. Low, 'Embodied Space(s): Anthropological Theories of Body, Space, and Culture' *Space and Culture 6.* 4 (2003): 10.

2 Shelley Smith and Nicolai Steinø, 'Public Space Design between Alienation and Appropriation the Case of Parkour,' in this volume.

3 Maurice M. Ponty, *Phenomenology of Perception,* trans. Forrest Williams (London, New York: Routledge, 2005), 64.

4 Alva Noe, *Action in Perception* (Cambridge, Massachusetts, London England: The MIT Press, 2004), 1.

5 Michel de Certeau, *The Practice of Everyday Life,* trans. Steven Rendall (Berkeley, Los Angeles, London: California Press, 1988), 117.

6 Angela Philip, 'Lost in Space,' *Back to the City, Strategies for Informal Urban Interventions,* eds. Angela Philip, Joanne Cys, Stephen Garret, Steffen Lehman (Berlin: Hatje Cantz Pub, 2009), 38–47.

invested in a way of acting, an art of combination which cannot be dissociated from an art of using.'[7]

De Certeau also describes how citizens move through an urban environment without seeing it.[8] 'Imaginary totalizations' produced by the eye create a kind of 'blindness' in the city.[9] However, by emphasising a 'two-way encounter between mind and city,'[10] a reflexive bodily experience occurs, which destroys daily routines and creates the body's own everyday practices in the city. This is how spontaneous, unexpected or strange representations of daily life become visible. It is not a known representation that is grasped from an urban text and seen all at once. Rather, it is a re-structured representation that is formed by the body and its experience in the city. Lavrinec explains this bodily transformation as a mutation from 'blind walker' to 'urban curator.'[11] The 'urban curator' is an active entity that comes in contact with its surroundings, becomes aware of it and is restructured by his or her experience. It distorts daily routines through new perceptions and continually highlights new spatial potentials. Bodies are no longer simply passive consumers, but are active participants who assist in the creation, construction and development of the city.[12]

Spatial configurations trigger, initiate and shape bodily experience in the city. The main focus of this study is on the participatory, experimental and temporary spatial interventions that activate the body in urban spaces. They are seen as having important potential for the creation of alternative ways of seeing or operating in the city.

We are currently facing a paradigm shift in the field of art and architecture. Rather than creating a simple, viewable object, both of these disciplines are searching for new approaches to the design of spatial installations that encourage interaction with the user. In the installations found in urban spaces, participants become the subject rather than simply having direct contact with the object. Also, Smith and Steinø[13] state that excluded groups and practices

7 De Certeau, *The Practice of Everybody Life,* 96.

8 Jekaterina Lavrinec, 'From a Blind Walker to an Urban Curator: Initiating Emotionally Moving Situations in Public Space,' *LIMES*: *Cultural rationalistic* 4.1 (2011): 54–63.

9 Michel de Certeau, 'Introduction: General Introduction.' In *The Practice of Everyday Life,* by Michel de Certeau (Berkeley, Los Angeles, London: California Press, 1988), IX–XV.

10 Michael Sheringham, 'City Space, Mental Space, Poetic Space: Paris in Breton, Benjamin and Reda,' *Parisian Field*, ed. Michael Sheringham (London: Reaction Books, 1996), 85–114.

11 Lavrinec, 'From a Blind Walker to an Urban Curator: Initiating Emotionally Moving Situations in Public Space,' 54.

12 Katerina Mojanchevska, 'Self-Organized Citizen's groups and Urban Space: Challenge in Planning Paradigm,' in this volume.

13 Smith and Steinø, in this volume.

are activating the public spaces by finding alternate routes and imaginary solutions. In this way the space becomes communicative. Participants can enable the space to become dynamic, reflective and communicative through the use of their bodies. Participants attach themselves to the urban space like prosthesis and make it a collective space through their own input. Similar to Katsuhiko Muramoto's[14] conclusions drawn from Merleau-Ponty's notions of the body and space in this volume, cities today are no longer a collection of passive infrastructure but are active environments that incorporate the activity of the body as a way to communicate. By disturbing the routine choreography of the city and in connection with Lima e Ferreira's[15] Theatre of the Oppressed and Mojanchevska's[16] Citizens' Groups, new art and architectural installations provide provocative mediums in which urban citizens can communicate with their surroundings and re-structure their own daily practices and bodily experiences.

This chapter will focus on art and architectural installations in public spaces that are unexpected, creative and temporary with the aim of discovering and describing their contribution to urban space. Body/space relationships will be placed at the forefront of the investigation. The following questions will be considered: How can we use our bodies as a tool for gaining experience and knowledge when we interact with our environment? What is the potential contribution of art and architectural installations in this communicative process? What kinds of mediums do they create in urban space? How do they enhance our spatial awareness and the performance of our bodies? How do they shape our cities as spaces for events? These questions will be explored by investigation of the selected examples, considering the strategy of the applied design of created spaces and the spatial tools used to transform the public space.

2 Spatial Interventions in Public Space

The term 'installation' is a relatively new term that is not found in dictionaries of fine art. In the 1950s, artists like Alan Kaprow used the term 'environment' to describe his work. In the 1970s, the terms 'project art' and 'temporary art' were used. It wasn't until the 1990s that the term 'installation' came into

---

14 Katsuhiko Muramoto, 'In Search of a Relational Body: Reflections on Sustainable Architecture,' in this volume.

15 Ana Carolina Lima e Ferreira, ' The Urban Challenge of Participation through Theatre in Rio,' in this volume.

16 Mojanchevska, in this volume.

use.[17] Contemporary, site-specific events, sculptural works and conceptual and performance art projects were early influences resulting in a major movement in the field of art. According to Sarah Bonnemaison and Ronit Eisenbach, the notion of 'installation' is critical because it demonstrates that art can be found outside of museums and galleries and can be a part of public space, created by public artists.[18] Art, in other words, had moved from static, permanent works to a more dynamic and temporary form giving people the chance to actively engage with their environment.[19] The works gave them tools for communication and interaction. The same characteristics provided inspiration for architects who have followed the lead of artists from the 1970s. They have also used their work to explore and interpret the nature of public space.[20]

Art and architectural installations have similar characteristics influencing one another in their refreshing approach to public space. An art installation can be defined as a three-dimensional work that is designed in relation to a specific space, location or situation. The characteristic of contextuality, or site-specificity, is the first aspect that merges the two disciplines. Secondly, both disciplines present experimental works with the aim of exploring new ideas through a dialog with the public. They attempt to create a place or an atmosphere or a situation where the creative imagination of both designer and the spectator can flourish. They trigger the body's own imagination by suggesting new possibilities for social contact in the urban space.

Thirdly, the focus of both disciplines is to provide people not with 'objects to view' but with 'spaces for action' where they can touch and experiment.[21] According to Alberto Perez-Gomez,[22] architectural design is usually identified with the production of novel and striking visual images. However, the experience of architecture is never merely spatial. Architectural design cannot be interpreted only as a manipulation of geometric spatial concepts. Indeed, our lived world is rich in sensation and emotions that arise from our bodily actions and engagements in the world. Along with Merleau-Ponty and Noe,

17 Sarah Bonnemaison and Ronit Eisenbach, *Installations by Architects, Experiments in Building and Design* (London: Princeton Architectural Press, 2009), 17.

18 Bonnemaison and Eisenbach, *Installations by Architects, Experiments in Building and Design,* 145.

19 Lima e Ferreira, in this volume.

20 Bonnemaison and Eisenbach, *Installations by Architects, Experiments in Building and Design,* 145.

21 Gray Read, 'Introduction: The Play's the Thing,' *Architecture as a Performing Art,* eds. Marcia Feuerstein and Gray Read (USA: Ashgate, 2013), 1–14.

22 Alberto Perez-Gomez, 'Two Analogical Reflections,' *in Architecture as a Performing Art* eds. Marcia Feuerstein and Gray Read (USA: Ashgate, 2013), 15–26.

Perez-Gomez indicates that 'perception is never purely passive reception; it is action.'[23] The motion of our embodied consciousness implies time. It is surprising to see that Karen Bradley, in his book *Rodolf Laban*, talks about the same idea focusing on the works of a dance artist. For Laban, 'space is the container for effort and shape and architecture is for action.'[24] In the world of architectural debate, this kind of spatial understanding reminds us of the concepts of 'event-space' or 'event cities' that was developed by Bernard Tschumi[25] in the 1990s. According to Tschumi, there is no space without event. Art and architectural installations in public space underline an 'action-oriented shift in thinking and making'[26] and enhance 'lived spatiality.'[27] By activating the body in the space, these spatial works speak with bodies and spectators. As with Muramoto's[28] examination of space, these are not objects but, instead, living spaces that exist only by the presence of bodies in action and communication. The conversation of the body with its surroundings emphasises the concept of 'participation through experimentation' as the fourth aspect of interest. Here we are talking about installation works in which the visitor is not a detached viewer but, instead, an active participant experiencing the space through his or her body. The body's movement does not end up with just a visual awareness but also a physical awareness in relation to space. In line with the works of artists Robert Morris, Trisha Brown and Bruce Nauman,[29] the architectural firm of Höweler & Yoon talks about developing participatory environments, interactive architecture and public space projects that combine architecture, media and landscape as new modes of public practice in the contemporary city.[30] The group is mostly interested in using networked technologies and new materials to set up urban environments that are inviting for the individual and call out for collective participation. They introduce the concept of 'play,' in

23 Ibid., 15.

24 Karen K. Bradley, *Rudolf Laban, Routledge Performance Practitioners* (USA: Routledge, 2009), 81.

25 Bernard Tschumi, *Event Cities 2, Praxis* (Cambridge: The MIT Press, 2001), 11–13.

26 Read, 'Introduction: The Play's the Thing,' 5.

27 Perez-Gomez, 'Two Analogical Reflections,' 15.

28 Muramoto, in this volume.

29 Stephanie Rosenthal, 'Choreographing You, Choreographies in the Visual Arts,' *Move, Choreographing You, Art and Dance since the 1960s,* ed. Stephanie Rosenthal (Cambridge, Massachusetts: The MIT Press, 2011), 8–21.

30 Meejin J. Yoon, 'Project at Play: Public Works,' *Back to the City, Strategies for Informal Urban Interventions,* eds. Angela Philip, Joanne Cys, Stephen Garret, Steffen Lehman (Berlin: Hatje Cantz Pub, 2009), 70–76.

other words, a state of being active or operational. They propose the construction of spatial environments where the body is invited to play in the city.

The fifth aspect being highlighted here is the notion of 'temporality.' The concept of temporality implies that spatial configurations change and evolve over time. Although Vitruvius emphasises durability and continuity as key concepts of architecture, Tschumi talks about a new kind of architecture that has to do with activating space through the movement of the body.[31] As it is also illuminated in Clementine Monro's[32] examination of her art installation and the ruin in this volume, concentration on the body/space relationship results in a paradigm shift from the stable and enduring architectural object to the temporary, impermanent, provisional, ephemeral and dynamic space in flow. This temporal characteristic of space opens up 'possibilities for a performative approach to architecture as a practice (a doing) and as a production (a thing done).'[33] Both temporal and spatial practices provide architects with broad freedom to create their design ideas in the public space and invite citizens to touch, experience and think about their surroundings. They also help to refresh an urban area infusing it with energy and, so, the concept of a 'pop-up city' is introduced.[34] Creative spatial interventions emphasise again the role of citizen participation in making cities. In a collective process, anyone can be a city-maker actively involved in the development of public space in the city.

The art and architectural installations that is the subject of this chapter create interactive mediums by which the urban citizen communicates with his or her surroundings, thus enhancing spatial awareness. With the help of these spatial apparatuses, the citizen re-structures his or her spatial perception in the midst of these spatial experiences. These interactive and relational structures render architectural space more sustainable, more sensitive and more conducive to individual involvement. In this way, and in line with the analysis of interventions of the body on space by Smith, Steinø, Lima e Ferreira and Mojanchevska in this volume, they encourage a new trend that moves from old, static, definite and finished icons in design towards more open, alive, changing, and responsive spatial entities. Here, the body becomes more productive and performative.

31 Omar Khan and Dorita Hannah, 'Performance / Architecture, An Interview with Bernard Tschumi.' *Journal of Architectural Education* 61.4 (2008): 52–58.

32 Clementine Monro, 'Perpetual Transitions: A Communication Between Photography and Absence Within the Spatial Experience of the Ruin,' in this volume.

33 Dorito Hannah, 'City as Event Space: Defying all Calculations,' *Back to the City: Strategies for Informal Urban Interventions*, ed. Steffen Lehmann (Berlin: Hatje Cantz, 2009), 114–119.

34 Jeroen Beekmans and Joop de Boer, *Pop-Up City: City Making in a Fluid World* (Amsterdam: BIS Publishers, 2014), 263.

It is suggested that these spatial interventions provide attempts for the development of public space by means of re-evaluating existing spatial potentials or revitalising them with new living scenarios. They also help architects and designers to create new spatial awareness through the process of thinking and making space in which the body becomes not an observer but an active generator. Creativity, productivity, performativity, collectivity, interactivity, temporality, flexibility, adaptability, spontaneity and site-specificity appear to be important design parameters in the process. The next part of the study concentrates on some examples of this type of spatial intervention. It attempts to explore the design tactics and their end products in terms of spatial configurations as well as the tools that are most often involved in the design process.

3 Decoding Art and Architectural Installations through Tactical Actions in Public Space

Public space is a place where citizens can express themselves, where they can interact and intervene with other places by creating their own space. Body and space relationships require a state of belonging as well. The body itself rebuilds the space – limiting, determining and directing its motion with body movements. 'I am the space where I am,' says the poet Noel Arnoud.[35] With this statement, he tries to explain the effect of body receivers, that is, the effect that the senses have in perceiving and experiencing the space. The body rebuilds its own space by creating its own perceptions. Even though the space is stable, the body and its awareness change according to particular conditions. As Smith and Steinø[36] have stated in their work, in public space, in order to create the space it desires, the body tries to create new ways to transform spatial perception. This part of the study deals with tactical actions in design, which enables the body to create new ways of negotiating with its surroundings.

A design tactic can be defined as set of rules and relationships or 'operative devices' that help designers to develop their design ideas.[37] They set out the ways in which the designer engages with his or her design problem by creating a dialogue with the space and proposing his or her own personal solution. The terminology draws inspiration from the term 'tactic,' which was used by De Certeau in his book, *Practice of Daily Life* in 1998. According to De Certeau,

35 In Gaston Bachelard, *The Poetics of Space*.(Boston: Beacon Press, 1969). 137.

36 Smith and Steinø, in this volume.

37 Manuel Gausa, 'City Sense: Territorializing Information,' *City Sense, Shaping Our Environment with Real-Time Data*, ed. IAAC (Barcelona, New York: Actar, 2012), 6–13.

tactics are 'proactive response to the city.'[38] In other words, 'they are practices that unfold through the dimension of time to transcend the spatial limits imposed by the powerful.'[39] Daily life creates its own 'operational tactics,' in order to deal with the strategies of power that impact the shape of the space. According to De Certeau, tactics involve calculations without their own space. Tactics lack a base where they can capitalise on advantages, prepare for expansion and ensure their independence from circumstances. These tactics produced by surprise are time-based and are driven by opportunism:[40] '[a] tactic depends on time; it is always on the watch for opportunities that must be seized on the wing. Whatever it wins, it does not keep. It must constantly manipulate events in order to turn them into opportunities.'[41] Tactics are the ways in which ordinary people engage with their surroundings and resist dictated spatial order. They are the public's way of 'knowing, thinking and acting' in the city.[42]

The design examples discussed in this chapter seek to serve public space by using different tactics such as creating new conceptual ideas, breaking boundaries without strict rules and re-defining the public space. These tactics are accepted as re-evaluated and adapted actions that aim to let society recreate their own space. These examples consist of collective spatial interventions that allow for the participation of every individual body. These spatial interventions try to interact with the viewer and make unexpected encounters possible as part of the daily routine of the body.

Fourteen examples have been chosen according to their degree of created spatial awareness and their contributions to public space. The spatial interventions are analysed in terms of the design tactics and spatial tools that have been used in their design processes. The first example is *Dispatchwork*.[43] In 2008 the street artist Jan Vormann began to repair cracked urban spaces using Lego® bricks. Vormann's Dispatchwork project became a remarkable spatial intervention that created social media connections all over the world. Later,

---

38 Janet McGaw, 'Performative Spatial Practices in the Urban Realm: A "Tactic" for Transcendence, Time Transcendence Performance,' *Conference Proceeding* (2010), 4, artsonline.monash.edu.au/performance/files/2012/09/proceedings-mcgaw-performative-spatial-practices-in-the-urban-realm-ttp-conference1.pdf.

39 Lain Borden, Joe Kerr and Jane Rendell eds., *The Unknown City, Contesting Architecture and Social Space* (Cambridge, Massachusetts, London, England: The MIT Press, 2000), 13.

40 De Certeau, *The Practice of Everybody Life*, 10.

41 De Certeau, 'Introduction,' ix.

42 Borden, Kerr and Rendell eds., *The Unknown City, Contesting Architecture and Social Space* 13.

43 'manifesto,' dispatchwork, accessed on date 03 May 2015, http://www.dispatchwork.info/manifesto.

Vormann's small-scale movement turned into an urban hacking movement. *Dispatchwork* emerged from Vormann's question: 'What are we not allowed to do to make the city better for ourselves?'[44] This project creates a new approach to the problem of broken walls through the use of a new construction material – Lego® bricks, a material that also adds colour to the grey urban landscape. Instead of being a single work, it becomes a work in which anyone can get involved and participate. Participants are ordinary people who are questioning the routine of daily life and want to add some playfulness to their experience of the city. Legos® emerge as toys that every citizen can play with and use to create a spatial medium that pushes the boundaries of the city. This characteristic of spatial intervention also emphasises the concept of the 'pop-up city.' By creating a playful experience in the city, it becomes possible for damaged buildings in the city to be repaired quickly with a new patching surface. This resonates with Smith's and Steinø's work,[45] where they argue that the certain rules in architecture control the space and appear as 'guilt party' of design. But when the citizens involve themselves with the public space and use the dynamics of the city itself freedom is created in bodily movement and expression.

The second example is Florian Rivière's *Trash Ball*.[46] An urban interventionist, Riviere turns city life into a playground by placing paint alongside trashcans in the street. An emerging layout is transformed into a design tool and urban surfaces are re-created using the idea of 'pop-up city.' Passers-by can dump their trash there. This situation increases the interaction between the body and public space and makes the individual a researcher and a creator of urban space. This urban hacking project brings a little playfulness into the issue of garbage collecting and proves how the behaviour of an individual in a public space can be changed when doing things becomes enjoyable. Transforming a *Trash Ball* into a playground area is an effective way to create environmental consciousness.

The third example is called *Stairway Cinema*.[47] It was designed to improve the poor quality of an urban situation by utilising existing spatial potential. Its purpose was to address the problem of lack of communication and Internet

44 Ibid.

45 Smith and Steinø, in this volume.

46 Bernard De Koven, 'Trashball and Other Works of Urban Hacktivism,' Deep Fun with Bernard De Koven, 04 May, 2012, Accessed on date 04 April 2015, http://www.deepfun.com/trashball-and-other-works-of-urban-hacktivism/.

47 'Stairway Cinema,' Oh.No.Sumo, Accessed on date 15 March 2014, http://ohnosumo.com/post/23707911789/stairway-cinema-stairway-cinema-is-the.

addiction and to increase personal contact by providing an alternative to the habit of checking cell phones while waiting at bus stops or corners. This mobile parasite invites passers-by to sit on cushions placed on stairways giving them the opportunity to watch short films alongside a busy Auckland Street. The program *Stairway Cinema* is organized by the audience itself – they control the installation. The created spatial structure consists of a timber frame covered with red fabric, which provides a real cinema experience. The result is an open, inviting and comfortable space where passers-by can stop for a few minutes and not use their cell phones. It serves as an experience that enriches the quality of structural function. The project aims to increase public awareness and the ability to pay attention to daily objects so that we do not become slaves of them. In the coming decades urban solutions like these will be a foundational component of a more sustainable and healthier form of urban living.

The fourth example is an installation called *Ready. Steady. Go.*[48] created by architects Sandra Janser and Elisabeth Koller. This unique installation creates a visual frame within the Jakomini area of Graz, Austria. Their idea was to paint the streets in the city of Graz in the form of a running track. The project invites the users and passers-by to experience the street in a new way. It is an urban renewal project that creates a new visual frame for these streets, giving visitors a fresh and inviting way to explore the city. The most important characteristic of this installation is the way in which it draws attention to and changes the usual round of daily life. It makes people stop and think: Why should we be in hurry all the time? Why are we behaving like runners as we go about our daily routine? The installation is questioning the daily routine, the feeling of always being in a hurry. The *Ready. Steady. Go.* installation is a very simple design idea that re-defines public surfaces – it is a thought-provoking example of created spatial identity.

The fifth example is a portable reading room called *Uni.*[49] The first *Uni* installation was in Manhattan. From New York it moved to Boston and Kazakhstan. The design idea emerged from a question: How can books be made more accessible in urban neighbourhoods? *Uni*, a mobile parasite, was developed as a way of sharing books, thus universalising the act of learning and improving interaction in the urban setting. It is a lightweight, portable structure consisting of two parts: a sitting area and bookshelves. The *Uni* project provides a

48 'Ready, Steady, Go.,' Graz City, Accessed on date 20 April, 2013, http://www.grazcityofdesign.at/en/output/detail/51/ready-steady-go.

49 'Uni, Portable Reading Room,' kickstarter, accessed 25 April 2014, https://www.kickstarter.com/projects/uni/the-uni-a-portable-open-air-reading-room-for-public (site discontinued).

spontaneous street-level experience that reconnects bodies with books. The designer team, called Street Lab, explains their project as follows:

> What we see at street level in many urban neighbourhoods does not reflect our aspirations for our society and ourselves. If we're serious about having a well-educated society, let's build cities where learning experiences are prominent, accessible and enjoyable. Let's show off our best teachers, librarians and educators doing great work and give them opportunities to adapt their craft to a public setting. The Uni takes learning public.[50]

Cities need new spatial solutions that are lighter-weight, more flexible, less expensive and better integrated into our patterns of daily life. As one example of this, *Uni* strengthens public memory, reveals city dynamics and enhances public dialogue.

When we look carefully and closely at our living environment we can easily recognize micro-level interventions in public spaces. These interventions could be characterized as 'city makers.' They are using the streets of the city as mediums for the construction of public criticism and the exploration of new conceptual ideas. The sixth example is from the Springtime Team and is called *Person Parking*.[51] It uses a design tactic that aims to return public space to public life. The project gives a parking spot, an area of two meters by five metres, back to the public by redefining the spot with the addition of furniture that allows bodily interaction and communication. This is what the public should have already. The aim of this mobile parasite is to improve the quality of public space by enabling people to enjoy themselves on streets that have been mostly dedicated to vehicles. Here the designers once again want to re-conceptualise public memory and the daily routines of the city. This spatial intervention, an urban hacking project, helps citizens to question the notion of parking spots in cities.

The 'idea of play' in urban public spaces aims to amuse people with accessible entertainment in the urban outdoors. Outdoor playing can take place at bus stops, bandstands and parks. Surfaces, walls and pavement can be played on in the moment or they can be watched remotely. The idea in creating a

50 Adale Peters, 'Street Smarts: The Uni Project's Outdoor Neighbourhood Reading Room,' Good, 14 February 2013, Accessed on date 15 March 2015, http://magazine.good.is/articles/street-smarts-the-uni-project-s-outdoor-neighborhood-reading-room.

51 'Person Parking,' *Yanko Design*, May 19, 2009, Accessed on date 20 April 2013, http://www.yankodesign.com/2009/05/19/park-it-mister/.

playing space is to initiate a collective movement and to organise unexpected conditions in the urban fabric. The seventh example concerns playgrounds. Here an open public space is transformed into playground, not for children but for adults. The *Off-Ground Installation*[52] is in Carlsberg City, Copenhagen. It consists of adjustable seating elements made of steel pipes and recycled fire hoses. Each seating element can be easily shifted to form a low seat, a hammock or a swing depending on the need of the user. Users can hang, swing or lay down – one size fits all. Through a re-conceptualising of one part of our daily routine this project reveals a new approach to the way that the citizen perceives and uses public space. The idea of play is mainly associated with children with the elements of play sized to fit the child. This project, however, reminds us that adults also need to play. The idea of playing makes public spaces less serious and thereby more enjoyable for the user. These spaces become lively mediums for the use and engagement of every citizen.

Another example is *White Noise, White Light,*[53] an interactive public space installation in Athens that was created by the architectural firm of Höweler & Yoon. It is a fifteen square meter area that has a view of the Acropolis. It consists of fibre optics and speakers. The creator of the project team, Höweler & Yoon, explains the genesis of the project as follows:

> There are so many sounds in the city you hear unadulterated, I didn't want to just record the city and play it back without filtering it in some way. We wanted to achieve something that was poetic, quiet and powerful all at the same time.[54]

White Noise consists of sound frequencies that are heard by everyone equally, *White Light* contains all colours of light. Sound receivers in the created field depress the sounds of the crowded city. Every stalk in the design unit is programmed with different algorithms that perceive body movement. When they perceive movement, white light shines brighter and the volume of the white noise gets louder. Conversely, when there is no movement around the stalks of the design unit, the light and noise decrease. As long as people pass through the field the design units observe them and provide representations of body

52 'Off-ground: Adults at Play,' *installation mag*, Accessed on date 15 March 2014, http://installationmag.com/adults-at-play/.

53 'White Noise, White Light,' *My Studio*, Accessed on date 04 April 2015, http://www.mystudio.us/projects/3.

54 Elizabeth Donoff, 'White Noise, White Light,' *Architectural Lighting*, Accessed on date 04 April 2015, http://www.archlighting.com/architects/white-noise--white-light.aspx.

motion. With motion, light and sound harmony the project creates a fluid and unexpected environment set against the view of the Acropolis.

As Wendell Pierce says, 'our thoughts are to the individual as our art is to the community.'[55] Accordingly, our next example is one that includes an individual's desires written on a public wall, thus becoming a work of art. The installation, *Before I Die*,[56] is a pop-up intervention that is being practiced on many public walls. After the loss of someone she loved, artist Candy Chang painted the side of an abandoned house in her neighbourhood in New Orleans. She wanted to create a daily reminder to all persons to take the time to consider what was most important to them. She stencilled this sentence on the wall: 'Before I die I want to____.' Anyone walking by can pick up a piece of chalk and reflect on their lives and their personal aspirations on this created wall. Over 550 *Before I Die* walls have been created in over 35 languages. Regularly thinking about death is a powerful tool for the rebuilding of perspective and reminds us of the things that make our lives meaningful. These walls are filled with fears, desires, hope and humour. Although it is a singular experiment, the project has drawn global attention. The design tool puts another layer on an abandoned wall. The public has created its own tactic for communicating with the rest of the public. This project indicates that our public spaces can help us make sense of the beauty and tragedy of life. It allows freedom of expression in a public space and enables those expressions to be viewed by others. Similar to Lima e Ferreira's[57] Theatre of the Oppressed, in this new relationship between body and space the individual finds an opportunity to share his or her desires and wishes by interacting with other people. This work increases the usage of public space, interrupts the daily routine, utilises existing spatial potential and carries the individual's experience into a common platform that transcends culture.

When you insert the 'idea of play' into the city life and create playful urban initiatives and interactions positive behavioural change can be seen in the community. *Noodle Delirium*[58] is a playful urban intervention created by Canadian designers, Les Astronautes. The project consists of pink and orange pool noodles placed on the facades of a narrow street. At the time, the street was unremarkable, forgotten. This project redefined an urban surface inviting passers-by to explore a new created space. The design team, Les Astronautes, describes the work as follows:

---

55 'Art,' *Candy Chang*, Accessed on date 06 April 2015, http://candychang.com/on-art/.

56 'Before I die,' *Before I die*, Accessed on date 10 March 2015, http://beforeidie.cc/site/.

57 Lima e Ferreira, in this volume.

58 'The Unusual Passages-Outdoor Art in Quebec City,' *Justin+Lauren website*, accessed on date 05 April 2015, http://justinpluslauren.com/unusual-passages-outdoor-art-in-quebec-city/.

> The installation transforms a discreet anonymous crack into an intriguing, colourful, textured passage. The stroller who ventures through it finds himself caught in a canal where pool noodles are used to modulate tactile, nearly organic paths that feel sensuous to some yet troubling to others.[59]

The next example belongs to Basurama, a well-known Spanish group. The designers turned the ruins of an abandoned electric train project in Lima, Peru into an amusement park. The space was recovered from the space set aside for the elevated electric railway. The project is called *The Ghost Train Park*.[60] Designers tried to create an environmental sensitivity that focused on the material waste of consumer society. Playing fields in *The Ghost Train Park* were built using recycled materials like car tires to build canopies, swings and basic climbing structures. During the construction period, the designers asked interested residents about their ideas about how to use this abandoned space. From these conversations emerged a collective team of designers, residents and volunteers who worked together to build a free recycled playground area for everyone. This project helped to ignite excited discussions on the management of public space and unused infrastructure with the needs, desires and creativity of urban residents emerging in the public sphere. Instead of demolishing existing derelict space, *Ghost Train Park* creators aimed to interfere and transform existing spatial potentiality, re-activating unused urban land through the use of recycled waste materials. They used old infrastructure as a part of the new design. Citizen participation in urban planning brings the real needs and desires of the citizen into the emerging space. Use of recycling materials, at the same time, leads to more economical and sustainable design.

The use of information and communication technologies, such as the Internet, as part of the design process is increasingly a part of emerging architectural forms. Their aim is to construct intelligent entities that continuously interact and interchange resources with their environment and with the users in the city.[61] *D-Tower*[62] is a permanent public art work in the city of Doetinchem, in the eastern Netherlands. Using its own website, the work aims to map the emotions of Doetinchem's inhabitants in response to a questionnaire determining

59 Ibid.

60 Aurora Fernandez Per and Javier Mozas eds., *Strategy and Tactics in Public Space.* (Spain: a+t Architecture Publishers, 2011.), 136–141.

61 Gausa, 'City Sense: Territorializing Information,' 13.

62 'D-Tower,' *Arcspace*. 20 January, 2003. Accessed on date 15 March , 2015. http://www.arcspace.com/features/nox/d-tower/.

the intensity of their feelings of love, hate, happiness and fear.[63] Based on their answers, the tower turns the emotions of inhabitants into colours, indicating blue (for happiness), red (for love), green (for hate) or blue (for fear). This project makes the emotional lives of inhabitants visible by the use of technological tools, interactive systems of relationships and network entities. The tall, prefabricated plastic structure helps us to view the reactions of inhabitants to events occurring in daily life. It also transforms the idea of traditional public space where people come together and share their experience. This is a new communicative and interactive medium in which people do not communicate with each other directly and physically but, instead, digitally and indirectly. The project also provides equipment underneath the tower where residents of the city can leave personal messages about the landscapes shown on the site.

Urban interventions in public space arise from both the dynamics of the city and from the wishes and desires of the individual. Similar to the outcomes of Mojanchevska's approach, the critical question is: 'Is citizen's contribution properly valued?'[64] In Mojanchevska's[65] study, it is mentioned that citizens have the opportunity to use a number of channels and activities to participate in the creation of the communal life. In this way, these art and architectural installations create their own space and then begin to be performed again in many other cities. Another example of this type of 'parasitic space' is an installation called *Little Free Library*.[66] Stereotank designed this project on Manhattan sidewalks. It aims to create a more habitable environment and a new way for communication to occur between body and space. In this project, the body communicates with the public with the help of books. The structure consists of a wooden frame with a bookshelf. The motto of the project is 'Take a book, return a book.' It fosters a collective relationship and open-source sharing. *Little Free Library* is a pop-up city movement that helps citizens to create their own space.

The final revamp project that we examine is the *Pink Ghost*[67] installation in Paris, France, created by Peripheriques Architects. In this work, a boring public space is transformed into a lively spatial medium by the addition of urban furniture and the use of an existing surface, creating an art object and

63 'D-Tower,' *D-Tower,* Accessed on date 06 April 2015, http://www.d-toren.nl/site.

64 Mojanchevska, in this volume.

65 Ibid.

66 'Little Free Library,' *stereotank,* Accessed on date 15 March 2015, http://www.stereotank.com/Little-Free-Library.

67 'Pink Ghost,' *peripheriques architects*, Accessed on date 13 June 2014, http://www.peripheriques-architectes.com/pink-ghost.

a meeting place. A plaza, shaded by large trees was transformed into a large collective space. Before the installation came into being people would walk through the plaza but not spend time there because there was nothing for them to do. The new pink, portable installation made the plaza attractive and inviting for people as a place to spend time. The five tables and twenty chairs made out of pink plastic enable conversation and public communication. It is a new interpretation of a living room in a public context. The work invites people from the neighbourhood and those outside of it to sit and enjoy the area. According to Bonnemaison and Eisenbach, the fantastic nature of this work triggers the imagination suggesting new possibilities for social contact in the urban area.[68] All aforementioned design works and implemented design tactics and tools are presented in Figure 7.1.

4 Spatial Interventions and Public Space

According to Lavrinec, spatial installations in public space produce certain types of behaviour and emotional experience and are able to activate citizens and encourage them to take part in the construction of the urban environment. Urban art and architectural interventions are productive and creative mediums, which articulate urban experience by rearranging the 'emotional landscape' of the city.[69] By disturbing everyday practices and routine spatial choreographies, they create experimental, interactive and reflective mediums in which citizens can communicate with their surroundings and re-structure their spatial perceptions and conceptions of the city. With the help of this provocative spatial apparatus, de Certeau's 'blind walker'[70] transforms into a reflexive 'urban activist.'[71]

Spatial readings through selected art and architectural installations in Figure 7.1 show that they are inspiring formations for architects who have great responsibility for the shaping of urban space. By implementing different design tactics, utilising existing spatial potential, re-activating unused urban land, creating mobile parasites, re-defining urban surfaces, using the idea of play, emphasizing technology or fostering the idea of the pop-up city, the designers present different approaches to engagement with urban space.

68 Bonnemaison and Eisenbach, *Installations by Architects, Experiments in Building and Design,* 148.

69 Lavrinec, 'From a Blind Walker to an Urban Curator: Initiating Emotionally Moving Situations in Public Space,' 58.

70 De Certeau, *The Practice of Everybody Life,* 93.

71 Ibid., 61.

	WORKS	TACTICS							TOOLS		
		Using the idea of play	Re-defining urban/public surfaces	Creating mobile parazites	Re-activating unused urban-land	Urban hacking + pup-up city	Emphasizing technology	Utilizing existing spatial potential	Surface/s	Spatial structure/s	Furniture/s
1	 Dispatchwork	✓	✓			✓		✓	✓		
2	 " Trashball "	✓	✓			✓			✓		
3	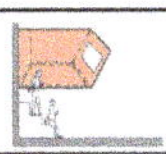 Stairway Cinema			✓		✓		✓		✓	
4	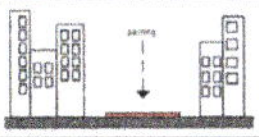 Ready. Steady.Go.		✓						✓		
5	 UNI, Urban Public Library			✓							✓
6	 Person Parking		✓	✓		✓		✓			✓
7	 Off - Ground Installation	✓									✓
8	 White Noise White Light						✓				✓
9	 'Before I die'		✓			✓		✓	✓		
10	 " Noodle delirium "	✓	✓		✓				✓		
11	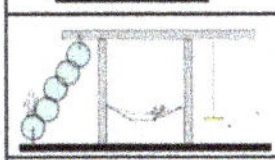 Ghost Train Park	✓			✓			✓			✓
12	 D - Tower						✓	✓			✓
13	 Little Free Library			✓		✓				✓	✓
14	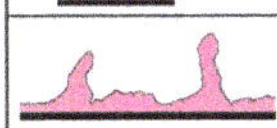 Pink Ghost		✓		✓			✓	✓		✓

FIGURE 7.1 *Design tactics and tools of the selected projects.*

They create playful furniture, interactive surfaces or small-scale spatial formations to re-define the relationship between body and urban space. This analysis demonstrates that temporary art and architectural installations are important resources with the potential to make significant contributions to city life. These installations manage to create a new public awareness by mirroring, revealing or complementing the unique characteristics of the chosen sites.[72] With the help of these spatial interventions, public space becomes an object that creates awareness in the daily life of the individual. The subject is now the individual who re-creates his or her personal space mentally, physically and emotionally.

Thierry Paquet talks about cities in his book, *Urban Bodies:*

> Is it possible to touch the cities? Yes, involuntarily (spontaneously), by falling down, walking through a wall or stumbling in front of a building. This action rarely occurs voluntarily. Although touching the city sometimes happens inherently, the individual wants to bring him/herself into existence and transform him/her in the space. He/she wants to create physical and emotional reactions by using the unsteady values of the location in which he/she exists.[73]

The art and architectural installations that are the subject of this chapter can be accepted as spatial tools by which the body can articulate itself in the city and create a reaction and interaction. In this manner, the body can touch the space and redefine spatial experience and its relationships with the urban space.

One of the ways we analysed the selected works in this chapter was with the examples of *Dispatchwork* and *Trash Ball,* which emphasize the idea of play using objects and materials that stick in the individual's memory. Here, the designers define the rules of the game and create possibilities for individuals to find their own way to interact or play with their surroundings. *Ghost Train Park,* on the other hand, emphasizes the idea of play by re-activating unused or abandoned urban land. By using existing spatial potentials the area is reclaimed into urban life through the participation of users. The *Off-Ground* installation interrupts the daily routines of adults in public space by

72 Steffen Lehman, 'Introduction: Hidden in the Urban Fabric: Art and Architecture, a Case Study of Collaboration in Interdisciplinary Context,' in *Back to the City, Strategies for Informal Urban Interventions,* eds. Angela Philip, Joanne Cys, Stephen Garret, Steffen Lehman (Berlin: Hatje Cantz Pub, 2009), 19.

73 Thierry Paquot, *Şehirsel Bedenler,* trans. Zeynep Bengü (İstanbul: Everest Pub, 2011), 17.

reminding them that the idea of play used to be a part of their daily life in childhood.

Society is a pattern of social relationships. This social pattern creates a spatial structure in cities and is most often invigorated in public space. *Stairway Cinema* addressed the loss of face-to-face physical interaction between bodies in urban space. It used the existing spatial potential and defined a mobile parasite spatial structure in which bodies could come together and watch a film. Another urban parasite is the installation, *Person Parking*. This example also aims to make use of existing potentiality by reclaiming city parking spots and creating a spatial platform for different activities such as sitting, reading, talking and resting. A similar approach can be seen in the work of the *Little Free Library*. This mobile parasite provides a simple spatial shelter, in which the body can find and read the books in urban space. This can be defined as an alternative urban library that encourages social dialogue. The *Uni* installation takes this idea a step further, introducing an open reading room with furniture, again making the experience of learning a pleasant and accessible activity for each citizen.

Public space creates a medium in which the freedom of expression of each individual becomes visible. Surfaces of urban spaces can be used as tools in this communication and expression process. Unused walls, floors or street spaces provide areas for bodily interventions. *Before I Die* is an installation that forms a communication wall on an unused urban surface. It starts with the idea of the reflection of individual feeling into an urban surface and then transforms into a collective work. In the *Pink Ghost* project, the designers again aim to reactivate a small scale unused urban square by filling the space with attractive furniture that creates a living room for the public. This strategy is also used on a larger street scale in the *Noodle Delirium* project.

Technology not only changes bodily communication and interaction in the space but also the ways of constructing spaces. *D-Tower* is an interactive urban installation that visualizes the emotions of city residents with the help of Internet technology. Similarly, *White Noise, White Light* creates an interactive installation, in which the movements of bodies are transformed into light and sound. This spatial intervention makes it possible for the body to re-structure urban space constantly through the use of the flow of movement.

5 Conclusion

Lane Borden, Joe Kerr and Jane Rendell argue that the city should bring together the micro-architectural and macro-planning scales, the everyday realm and the urban realm. The city must be situated between the perceived and the lived:[74] 'Architecture emerges not as an object, not as a thing but as a flow, the merely apparent pattern of a much more complex set of forces, dynamics and interrelations within the space of the city.'[75]

An important twenty-first century architect, Rem Koolhaas, states that the void is not something to fear but rather is a site for potential, a site for performance, a 'liberty zone.'[76] Metropolitan voids do not stay empty, they are appropriated for unexpected programs of activity. For Koolhaas, 'performative practices are generated by occupants, the role of architecture is to create space in which they can occur.'[77] Based on this statement, it is suggested that art and architectural installations provide productive spaces that trigger performative practices in urban space. Here, with these spatial creations the designer initiates the action but he or she does not control it. The one that keeps this action alive is the citizen, the owner of the public space, in other words, the individual him or herself.

Bibliography

'Art.' *Candy Chang*, Accessed on date 06 April 2015. http://candychang.com/on-art/.

'Before I die,' *Before I die*, Accessed on date 10 March 2015, http://beforeidie.cc/site/.

'D-Tower,' *Arcspace*. 20 January, 2003. Accessed on date 15 March 2015. http://www.arcspace.com/features/nox/d-tower/.

'D-Tower,' *D-Tower*, Accessed on date 06 April 2015, http://www.d-toren.nl/site.

'Little Free Library,' *Stereotank*, Accessed on 15 March 2015. http://www.stereotank.com/Little-Free-Library.

'Manifesto,' Dispatchwork, accessed on date 03 May 2015, http://www.dispatchwork.info/manifesto.

'Off-ground: Adults at Play.' *installation mag*. Accessed 15 March 2014. http://installationmag.com/adults-at-play/.

74 Borden, Kerr and Rendell eds., *The Unknown City, Contesting Architecture and Social Spac*, 20.

75 Ibid., 20.

76 Janet McGaw, 'Performative Spatial Practices in the Urban Realm: A "Tactic" for Transcendence, Time Transcendence Performance,' *Conference Proceedings* (2010), 1–17, artsonline.monash.edu.au/performance/files/2012/09/proceedings-mcgaw-performative-spatial-practices-in-the-urban-realm-ttp-conference1.pdf, 7.

77 Ibid., 7.

'Park it mister,' *Yanko Design*, May 19, 2009. Accessed April 20 2013. http://www.yankodesign.com/2009/05/19/park-it-mister/.

'Pink Ghost,' *Peripheriques architects*. Accessed on date 13 June 2014. http://www.peripheriques-architectes.com/pink-ghost.

'Ready, Steady, Go.' *Graz City*, Accessed on date 20 April 2013. http://www.grazcityofdesign.at/en/output/detail/51/ready-steady-go.

'The Unusual Passages-Outdoor Art in Quebec City,' *Justin+Lauren website*, accessed on date 05 April 2015. http://justinpluslauren.com/unusual-passages-outdoor-art-in-quebec-city/.

'Uni, Portable Reading Room.' *Kickstarter*. Accessed 25 April 2014. https://www.kickstarter.com/projects/uni/the-uni-a-portable-open-air-reading-room-for-public (site discontinued).

'White Noise, White Light.' *My Studio*. Accessed on date 04 April 2015. http://www.mystudio.us/projects/3.

Aurora Fernandez Per and Javier Mozas eds., *Strategy and Tactics in Public Space*. (Spain: a+t Architecture Publishers, 2011a), 136–141

Bachelard, Gaston. *The Poetics of Space*, Boston: Beacon Press, 1969.

Beekmans, Jeroen and Joop de Boer. *Pop-Up City: City Making in a Fluid World*. Amsterdam: BIS Publishers, 2014.

Bonnemaison, Sarah and Ronit Eisenbach. *Installations by Architects, Experiments in Building and Design*. 14–21. London: Princeton Architectural Press, 2009.

Bradley, Karen K. and Rudolf Laban, Routledge Performance Practitioners. USA: Routledge, 2009.

De Certeau, Michel. 'Introduction: General Introduction.' In *The Practice of Everyday Life*, by Michel de Certeau, xi–xv, 96–117. Berkeley, Los Angeles, London: California Press, 1988.

De Koven, Bernard. 'Trashball and Other Works of Urban Hacktivism.' Deep Fun with Bernard De Koven. 04 May, 2012. Accessed on date 04 April 2015. http://www.deepfun.com/trashball-and-other-works-of-urban-hacktivism/.

Donoff, Elizabeth. 'White Noise, White Light.' *Architectural Lighting*, Accessed on date 04 April 2015. http://www.archlighting.com/architects/white-noise--white-light.aspx.

Feuerstein, Marcia and Gray Read eds., Architecture as a Performing Art, 1–14. USA: Ashgate, 2013.

Gausa, Manuel. 'City Sense: Territorializing Information.' *City Sense, Shaping Our Eenvironment with Real-Time Data*, edited by IAAC, 6–13. Barcelona, New York: Actar, 2012.

Hannah, Dorito. 'City as Event Space: Defying all Calculations.' *Back to the City: Strategies for Informal Urban Interventions*, edited by Angela Philip, Joanne Cys, Stephen Garret, Steffen Lehman, 114–119. Berlin: Hatje Cantz, 2009.

Khan, Omar and Dorita Hannah. 'Performance / Architecture, An Interview with Bernard Tschumi,' *Journal of Architectural Education* 61.4 (2008): 52–58.

Lain, Borden Joe Kerr and Jane Rendell eds., *The Unknown City, Contesting Architecture and Social Space.* 13–20. Cambridge, Massachusetts, London, England: The MIT Press, 2000.

Lavrinec, Jekaterina. 'From a Blind Walker to an Urban Curator: Initiating Emotionally Moving Situations in Public Space.' *LIMES : Cultural Rationalistic* 4. 1 (2011): 54–63.

Lehman, Steffen. 'Introduction: Hidden in the Urban Fabric: Art and Architecture, a Case Study of Collaboration in Interdisciplinary Context.' *Back to the City, Strategies for Informal Urban Interventions.* edited by Angela Philip, Joanne Cys, Stephen Garret, Steffen Lehman. 14–35 Berlin: Hatje Cantz Pub, 2009.

Low, Setha, M. 'Embodied Space(s): Antropological Theories of Body, Space, and Culture.' *Space and Culture* 6.4 (2003): 10.

McGaw, Janet. 'Performative Spatial Practices in the Urban Realm: A "Tactic" for Transcendence,' Time Transcendence Performance,' *Conference Proceedings* (2010), 1–17, artsonline.monash.edu.au/performance/files/2012/09/proceedings-mcgaw-performative-spatial-practices-in-the-urban-realm-ttp-conference1.pdf.

Noe, Alva. *Action in Perception.* Cambridge: MIT Press, 2004.

Paquot, Thierry. *Şehirsel Bedenler* (*Urban Bodies*), translated by Zeynep Bengü. Istanbul: Everest Publisher, 2011.

Per, Aurora F. and Javier Mozas eds., *Strategy and Tactics in Public Space,* Spain: a+t Architecture Publishers, 2011b.

Perez-Gomez, Alberto. 'Architecture as a Performing Art: Two Analogical Reflection.' *Architecture as a Performing Art* edited by Marcia Feuerstein and Gray Read, 15–26. USA: Ashgate, 2013.

Peters, Adele. 'Street Smarts: The Uni Project's Outdoor Neighbourhood Reading Room.' Good. 14 February 2013. Accessed on date 15 March 2015. http://magazine.good.is/articles/street-smarts-the-uni-project-s-outdoor-neighborhood-reading-room.

Philip, Angela. 'Lost in Space.' *Back to the City, Strategies for Informal Urban Interventions,* edited by Angela Philip, Joanne Cys, Stephen Garret, Steffen Lehman 38–47. Berlin: Hatje Cantz Pub, 2009.

Ponty, M. Maurice. *Phenomenology of Perception.* Translated by. Forrest Williams. London: Routledge, 2005.

Read, Gray. 'Introduction: The Play's the Thing.' *Architecture as a Performing Art,* edited by Marcia Feuerstein and Gray Read, 1–14. USA: Ashgate, 2013.

Rosenthal, Stephanie. 'Choreographing You, Choreographies in the Visual Arts.' *Move, Choreographing You, Art and Dance since the 1960s.* edited by Stephanie Rosenthal. Cambridge, Massachusetts: The MIT Press, 2011.

Sheringham, Michael. 'City Space, Mental Space, Poetic Space: Paris in Breton, Benjamin and Reda.' *Parisian Field,* edited by Michael Sheringham. 85–114. London: Reaction Books, 1996.

'Stairway Cinema.' Oh.No.Sumo. Accessed March 15 2014. http://ohnosumo.com/post/23707911789/stairway-cinema-stairway-cinema-is-the.

Tschumi, Bernard. *Event Cities 2, Praxis,* Cambridge: The MIT Press, 2001.

Yoon, Meejin J. 'Project at Play: Public Works.' *Back to the City, Strategies for Informal Urban Interventions,* edited by Angela Philip, Joanne Cys, Stephen Garret, Steffen Lehman. 70–76. Berlin: Hatje Cantz Pub, 2009.

CHAPTER 8

Self-Organised Citizens' Groups and Urban Space: Challenges in Planning Paradigm

Katerina Mojanchevska

Abstract

The decades of transformation from government to governance have increasingly been about redistributing political power and citizens' participation in decision-making processes, both on national and local levels. The rationale behind this institutional redesign is the benefit of direct participation of those affected in the decision-making process. In the more immediate relationship with the citizens and the more fragmented sources of power, cities are able to democratise the public space from below and include diverse publics (and not only those skilled and vocal in their claims) into deliberation over redistributive process of public goods. Various theories of policy-making seek to encapsulate the new relation between the citizen(s) and the institutions, such as: collaborative policy-making,[1] communicative planning,[2] the Just City approach,[3] multicultural planning,[4] to name but a few. While being quite distinctive and drawing inspiration from different disciplines, they share some common features, such as disappointment with technocratic planning and beliefs that the civil society and self-organised citizens' groups are the key to social transformation and empowerment of groups outside (and sometimes against) the state.[5] Yet, there are serious challenges in meeting these objectives. Andrea Cornwall and Vera Coelho note that widening participation is more than 'invitations to participate' and 'for people to be able to exercise their political agency, they need first to recognise themselves as citizens rather

1 David E. Booher and Judith E. Innes, 'Network Power in Collaborative Planning,' *Journal of Planning Education and Research* 21.3 (2001): 221.

2 Patsy Healey, *Collaborative Planning: Shaping Places in Fragmented Societies* (Basingstoke: Macmillan, 1997).

3 Susan Fainstein, *The Just City* (New York: Cornell University Press, 2010).

4 Leonie Sandercock, 'When Strangers Become Neighbours: Managing Cities of Difference,' *Planning Theory and Practice* 1.1 (2001): 28.

5 Vanessa Watson, 'The Usefulness of Normative Planning Theories in the Context of sub-Saharan Africa,' *Planning Theory* 1.1 (2002): 29.

 | DOI 10.1163/9789004365520_010

than as beneficiaries or clients.'[6] In advocating participatory approaches that operate beyond representational democracy and voicing the perspectives of different social groups in respective bodies and spaces, this chapter studies if and in which ways do self-organised citizens' groups shift the power balance in urban and social making of cities and if the control over public planning process is real or symbolic. The research is based on qualitative methodology and is limited to the city of Skopje, Macedonia and its neighbourhoods.

1 Introduction

Since 2009 the city of Skopje, the capital of Macedonia, has been subject to an urban revitalisation project titled 'Skopje 2014,' which intends to result in a new urban imaginary. After almost two decades of transitional decay and urban planning subordinated to commercial forces, the government backed by the city authority, issued a visual design of the 'new' city and the urban core. The events surrounding this project inspired a genuine social mobilisation and an open confrontation with the governmental and political elites in the city.

This project polarised the public: on one side were those supporting the renewal of the city and on the other, those requesting a debate on what the new symbolic order of the city should be. It made clear that urban development in general and the planning and use of public space in particular had importance to the citizens. Within the years to follow, these issues gained importance as the institutional redesign and decentralisation initiated in 2004 did not produce the desired conviviality effects between governments and citizens. There is a lack of recognition and legitimisation of citizens' consent and involvement in public actions as well as a gap between the legal and technical side of institutionalising participation and the reality of effective citizens' involvement.[7] Macedonia is growing ever more divided and (the abuse of) public space planning is at the centre of this destabilising process.

Despite the evident shift in citizens' involvement in public policy making, as Boonstra and Boelens argue, participatory planning is still developed and administered by public government and the methods used to promote

6 Andrea Cornwall and Vera Schattan P. Coelho, Spaces for Change? The Politics of Participation in New Democratic Arenas.' *Spaces for Change? The Politics of Participation in New Democratic* Arena, eds. Andrea Cornwall and Vera Schattan P. Coelho (London: Zed Books, 2007), 8.

7 Cornwall and Coelho, 'Spaces for Change?' 3.

citizens' involvement are largely based on government's preconditions.[8] In such context, we can expect what Shelley Smith and Nikolai Steinø in this volume term as 'designated public space use' utilising 'formalised codes of accepted behaviour.'[9] Limiting citizens' involvement in imagining space, as Iklim Topaloğlu and Pelin Dursan Çebi[10] discuss in this volume note, takes away their active role in the process of city-making. As a result, both Smith and Steinø and Topaloğlu and Çebi, comment that urban planning is no longer contributing toward 'inclusive and stimulating urbanity'[11] and collective spaces are not made through citizens' input.[12]

This chapter intends to facilitate critical re-thinking of the urban space planning paradigm. Resonating with Ana Carolina Lima e Ferreira's[13] examination of Legislative Theatre in this volume, it suggests that designing public space is a powerful instrument of social control[14] and that the institutional designs that enhance participation in a formal way, such as electoral bodies and self-governance mechanisms, should be coupled by social mobilisation and genuine motivation and willingness of the citizens to participate in decision-making processes. In order to understand why and how citizens can and will contribute on the basis of self-motivation to urban development, it is important to look at the structure and operation of self-organised citizens' groups.

2 Mechanisms for Citizens' Direct Participation in Local Policy Making

Citizens have the opportunity to use a number of channels and activities to participate in the creation of the communal life. The critical questions are: '*Are citizens informed about them, and is their contribution is properly valued?*'

8 Beitske Boonstra and Luuk Boelens, 'Self-organization in urban development: towards a new perspective on spatial planning,' *Urban Research & Practice* 4.2 (2011): 99–100.

9 Shelley Smith and Nicolai Steinø, 'Public Space Design between Alienation and Appropriation the Case of Parkour,' in this volume.

10 Iklim Topaloğlu and Pelin Dursun Çebi, 'New Interventions in Public Space: Spatial Readings through Art and Architectural Installations,' in this volume.

11 Smith and Steinø, in this volume.

12 Topaloğlu and Çebi, in this volume.

13 Ana Carolina Lima e Ferreira, 'The Urban Challenge of Participation through Theatre in Rio,' in this volume.

14 Alexander R. Cuthbert, *The Form of Cities: Political Economy and Urban Design* (Oxford: Blackwell Publishing, 2006), 79.

Besides the conventional forms of communication between government and citizens, such as rally protests, civic initiatives and referendums, information technology development has created new opportunities and ways of communication and the Internet and its forms create a new public space for citizens to action. With the relatively high percentage of use of the Internet in Macedonia (68.1% in the first quarter in 2014 and 55.2% used mobile Internet or smart phone access outside of their office or home),[15] the potential of this tool is yet to be utilised.

Citizens have the constitutionally guaranteed right to local self-government which is exercised indirectly through representatives in municipal bodies, elected at legitimate elections and indirectly through the legal right of assembly and association. Direct participation of citizens in decision-making at the local level is organised as part of a civic initiative, a public meetings, referendums, giving suggestions and submissions and public hearings. These are so-called direct forms of democracy or immediate democracy. In general, all require that proposals and claims are submitted to the municipal bodies (the Council or the Mayor) and if legal pre-conditions are respected in relation to threshold percentage and appropriateness of the issue, then the Council or the municipal administration are obliged to have them discussed and file their response.

The most commonly used form of civic organisation in Macedonia is the civil-society organisation. Macedonia has a relatively high percentage of civil-organisations (5.769 organisations or 2.85 per 1.000 inhabitants) operating in diverse fields of action, as: social issues, youth, culture, employment health, etc. According to the analysis of the civil society sector conducted by Macedonian Centre for International Cooperation (MCIC) in 2005,[16] the civil society in Macedonia is well organised with a strong supporting infrastructure and a moderate networking level. As a result of lack of social capital and other social and cultural norms in this sector in the country, it is relatively limited in the width and depth of citizens' involvement. *Can self-organisation be an answer to these deficiencies?*

---

15 State Statistical Office of Republic of Macedonia, *Use of ICT in households and by individuals* (Koristenje na informatichko-komunikaciski tehnologii vo domakjinstvata i kaj poedinci), October 2014.

16 Saso Klekovski, Goran Buldioski, Fatmir Bitikji, Emina Nuredinoska, Natasa Gaber-Damjanovska, Kelmend Zajazi, Suncica Sazdovska, and Biljana Gerasimovska-Kitanovska, *Po 15 godini tranzicija – Od stabilizacija kon gragjansko uchestvo* (After 15 years of Transition – From Stability to Citizen's Participation) (Skopje: Macedonian Center for International Cooperation, 2006), 10–11.

In the absence of a solid, stable regulatory urban planning system, many Balkan cities today rely on vigorous self-organisation. The gap between formal and informal parts of urban development placed local self-organised citizens groups in a difficult position to mitigate between informal and formal planning and integrated vs. segregated areas into urban policy. Cornwall and Coelho note that widening participation is more than 'invitations to participate' and 'for people to be able to exercise their political agency, they need first to recognize themselves as citizens rather than as beneficiaries or clients.'[17] In such re-awakening, the human body's potential to influence public space transformation can be detected, one that Smith and Steinø[18] succinctly comment in this volume.

3 Self-Organised Groups of Citizens and City-Making

The simplest definition of a self-organised group, which goes beyond the complex system approaches, is one that binds citizens by self-motivation in autonomous and community-based networks that are outside of governmental control.[19] The definition accentuates their internal motivation and independency. However, the understanding of the relation between agency and structure is an important issue in the initiative's success and barriers.

The collaborative/interactive planning approaches advocated by Innes, Healey and Booher are increasingly criticised as working in favour of the formal urban planning, which marginalises the less powerful citizens.[20] They are regarded as time-consuming procedures that only a few people have time for. Moreover, governments are not always willing to collaborate, negotiate and change because of civic initiatives that challenge their views and policy decisions.[21] Governments and policy makers are promoting, or more precisely, *requiring* citizen involvement in planning everyday life. Yet, the process by which this is stimulated keeps the government as leader with decisive power. As such, citizens' involvement is no longer a shared responsibility between governments and citizens but rather a barrier in planning paradigm re-design.

---

17 Cornwall and Coelho, 'Spaces for Change?' 8.

18 Smith and Steinø, in this volume.

19 Boonstra and Boelens, 'Self-organization in urban development,' 100.

20 John Pløger, 'Strife – urban planning and agonism,' *Planning Theory* 3.1 (2004): 77.

21 Isabelle Doucet, 'Negotiating Brussels: collectives in search for a common world?' In: *Urban ontologies: importing ANT into the urban studies seminar*, (Berlin, Germany: Centre for Metropolitan Studies, 2007), 7.

The guiding questions of this research are: *How space in Skopje is produced, and which are the spaces of participation of citizens in the process of urban and social planning of the neighbourhoods through self-organisation? What are the strengths and obstacles of the local citizens' self-organised groups in the process of urban planning of their neighbourhoods? What is the power distribution between official planning and self-organised groups for urban development?*

In this chapter, self-organisation refers to active citizens and their endeavours in the neighbourhoods to protect the green areas and children's space, defend the public space and city squares and to support the less privileged co-residents of the city.

4 Methodological Framing

The research is based on qualitative methodology to grasp the perceptions, attitudes and actual behaviours of self-organised groups of citizens. It included semi-structured interviews organised in March–April 2014 with representatives of eight (8) groups active in the period 2009–2013 in Skopje whose activities relate to urban development and public space. Representatives of the following groups took part in the research:

- Self-organised group of citizens for protection of the public park in the municipality of Chair: the group was established in January 2013 when first rumours about building of residential blocks within the green areas of the park in Chair were spread among the residents. Taking into consideration the fact that the municipality of Chair is among the most densely populated areas in Skopje, with a territory of only 3.5 km^2 and a population of 64.773 residents, with few green areas in the urban part of the municipality, the protection of the park was even more significant. The group was organised by the residents from the buildings around the park, directly affected by the new urban plan. The group appointed a well know public figure as their leader and provided support from approximately 200 residents from the nearby buildings. Currently the group is not active but they are closely monitoring the situation related to the detailed urban plan for the area.
- Hrana, ne oruzje (Food no War): a self-organised group of young people, mostly artists and friends who share similar concepts for human transformation of the society in a creative way. They address the issue of urban poverty and through joint cooking and public sharing of food they try to bring comfort and hope for the hungry and homeless people in Skopje. The group is operational for three years with a core of 15 members.

- Self-organised group of citizens from the 'Meksichka' street: this is a group of residents at the 'Meksichka' street directly affected by the decision of the local government to allow building of a new residential block on the children's playground and park at that street. Moreover, the only monument reminding the citizens of Skopje about the solidarity of the Mexican people in rebuilding the ruined city in 1963 earthquake was located in that park. Therefore, the street was named Meksichka (The Mexican street). Recently the street name has been changed to symbolically alter the collective memory of its residents. The group has been active for one year with a core of five families living around the park and the playground.
- Prva Arhi-brigada – PAB (First Archi Brigade): a self-organised group of students of architecture, established in 2009 as a response to the urban changes and occupation of public space in the central city core with intention to provoke open public debate on the new urban imaginary of Skopje. In the beginning, there were 5 core members of the initiatives, but through their activities the number increased. PAB and Ploshtad Sloboda organised the First Architectural Uprising on 28th March 2009. Currently the group is not active. However, they are still debating the issue through their blog and Facebook page.
- Ploshtad Sloboda (Freedom Square): initiated as a self-organised group of urban and cultural activists which together with First Archi Brigade intended to open a public dialogue, inclusive for all stakeholders in relation to the urban planning of the city centre and the public space, in general. Later on, the group transformed into a civil-society organisation. The organisation is still active, mostly engaged in communal grass-root activities.
- Parkobrani (Park-defenders): a self-organised group of activists and residents from the area around hotel Bristol (located in the immediate centre of Skopje). The group protested against the destruction of the green area around their buildings and the hotel aimed to provide space for new administrative buildings for public companies. The growing tendency of the city administration to trim down the green areas in Skopje, especially in the urban core, and the rising problem with the air pollution, gave this action even greater importance. Currently the group is not active.
- Raspeani skopjani (Singing citizens of Skopje): a self-organised group of activists which stream from the core group of those involved in the events on 28th March 2009. Similar to Lima e Ferrera's[22] Theatre of the Oppressed, this city singing group is composed of around 10 people that use urban singing

22 Lima e Ferreira, in this volume.

as a performance tool to react to the urban and social happenings in Skopje, including the political constrain of the right to disagree.
- Spas za Vodno (Save Vodno): a self-organised group of hikers, mountain lovers and citizens of Skopje organised to save the nearby mountain Vodno from further infrastructural destruction and raise public attention to Vodno as 'the green lungs' of Skopje and a 'getaway' from the urban jungle. It was initiated in October 2013 with a core team of 10 people. However, the group decreased in size and activities, even though the importance of the issue is growing bigger and stronger.

A limitation of the study is its convenience sample that hinders generalisation of the findings to other contexts. Its importance lies in the fact that it is a pioneer in attempting to research the phenomena of 'self-organisation' in the context of urban development in the city of Skopje. It provides a skeleton upon which further knowledge can be built and criticised. The research adds the value of understanding the roles of the citizens and the public spaces in the social dynamics of the city. It adds to the body of literature interested in the civic function of public spaces as something that is a 'natural' part of the routine of everyday life and is both shaped and influences power struggles among diverse social groups.[23]

5 Findings

Group Formation and Internal Conflict Management

In general, the self-organised groups of citizens in Skopje have been formed spontaneously, either by members that knew each other well and shared similar personal and professional interest or by neighbours sharing similar needs of the immediate public space. One of the groups was formed by people that organised themselves via social media channels (Facebook). Others formed as a result of integration between resident activists who previously initiated some actions in the given area and a group of civic activists that worked on the same issue. Friendship is an important factor that makes the group cohesive. Development of close ties is an important outcome of the group's activity and when the group is unable to foster such relations, the number of its members shows tendency to decrease.

23 Tridib Banerjee, 'The Future of Public Space,' *Journal of the American Planning Association*. 67.1 (2001): 16–17.

The internal organisation is an important aspect for the group's operation. Groups that were aware of the capacities of their members and managed to delegate tasks were more effective in their work. Even in cases when their goal was not achieved, these groups have a sense that for their members it was a learning experience and maturation, on both personal level and community level.

Another important aspect for self-organised groups is internal conflicts management. Among the biggest impediment in the current context is the political affiliation of the group members. There is a tendency among the self-organised groups to speak for its members and initiatives as 'politically neutral.' The fear, or rather the stigma of being politically labelled is in a way understandable. Political parties and part of the citizenry perceive these forms of organisation as satellites of their political opponents without any civic value. Such perception drives citizens away from the initiative even when their personal ideas and views coincide with those of the group. In addition, very often the media enters into the spinning game of 'random' errors and incorrect representation, thus creating additional pressure on the group members to reconsider their independence and self-motivation. In many cases, in order to save the initiative from labelling and to mitigate internal conflicts, some of the groups asked their 'unsuitable' members to leave the group. This in itself is a negative tendency because the essence of self-organisation is based on the free will of individuals to act and equal respect for everyone's opinion. Such radical exclusory measures are therefore antithetical. However, it seems that in this current context, self-organised groups to some extent are a reflection of the divided society. These divisions are not motivated by economic, social, ethnic or gender differences but by their connections with any of the political parties. Political party affiliation is considered corrosive to the group cohesion and as something that does more harm than good for their mission. It trivialises their efforts and makes the problem look insignificant, irrelevant or, at the very least, not important enough to be reviewed by the authorities or the public as whole. The real challenge for self-organised groups in Skopje (and in Macedonia) is how to avoid political polarisation and instead keep the focus on the relevance of their legitimate demands.

Strategic or Tactical Action in Fulfilling the Objectives

The research identified two approaches used by the self-organised groups in terms of their efforts to achieve the goals and approach their target groups. One is the so-called strategic approach and the other is tactical. The focus of the first is on the actors and actions that originate from spontaneously developed strategies or coherent group behaviour. The other is more focused on actions and protests, such as swift response to the dynamic and changeable environment. It is fair to say that self-organisation in Skopje has evolved in response to the inability to find a satisfactory compromise between policy objectives and civic responses to the offered solutions. Moreover, because of the legal framework that allows for citizens to be consulted in the final stages of development of the plans, the expectation is that their responses to the new situation will be reactive rather than proactive. Therefore, legitimacy of the self-organised groups to demand change and participation in the planning stage is often questioned, as is the tension between their policy objectives and realisation through self-organisation.

Strategy and internal dynamics determines the ability of the group to reach the target groups and potential supporters. As a result, some are perceived as elite groups, while others like open platforms where one can freely choose to join or leave.

The basic question that every form of civic organisation should be able to answer is the type of relationship that they want to develop with the institutions. In this sense, it is important for the group to know whether their goal is to take part in the tasks of the institutions, to be complementary to their work or to have autonomy in their activity, etc. In the case of Skopje, self-organised groups mainly seek a complementary function to that of the existing institutional structures and also to serve as a consultative body when decisions are made about a particular area. They all respond reactively and were radical in their actions using protests, marches and occupations of public space and manifestos to express their views and the like, although, basically they all seek to establish a dialogue with the authorities and presented themselves as a partner for cooperation. Despite this cooperative spirit, the complex relationship between civil society and structures of power hampers the adaptability and communication between existing institutional structures and self-organised initiatives. In practice, this was achieved only in the case of the Chair. Only one self-organised group of citizens stated that their tactical approach was not to demonstrate cooperation but rather to show their rebellious and disobedient spirit.

It is important to understand which models of communication between the formal and informal groups enabled connectivity between these two sectors, especially seen in the context of the complexity of their relationships. The groups should act proactively and try to give their suggestions or alternative solutions to the current situation rather than just be reactive. Their openness to cooperation should be regarded as an expression of interest to participate and contribute to the decision-making process. However, that does not mean that they accept to be a décor who will give a carte blanche support to the decisions of the structures of power. But is this possible? As one correspondent stated, 'to talk about participation in the current context is nonsense. It is all about giving and obtaining and ignorance. If at all there is participation, then they (self-organised groups) serve only to fill in the quotas, only declaratively, to provide an alibi.'[24]

Groups' Leadership

Self-organised groups of citizens usually choose horizontal managerial structure. Only one group had a formal leader, a person that invested his personal capacities in obtaining wider public and media attention, so he was recognised as 'the face' of the initiative and the group was successful in meeting its objectives. Other groups utilised delegated tasks based on members' capacities. Decisions were jointly made but in some situations when fast reaction was needed it proved to be time consuming and obstructive. Still, they find this kind of operation as most suitable to their concept of civic engagement. Everything else, according to them, would lead to institutionalisation. In a third case, a group started within a flat structure but after a while the group number decreased and the person that felt the most attached to the initiative was recognised by the others as a leader, though she did not feel like it and did not want to fight for the idea by herself. 'I truly identify myself with the idea but I didn't get involved in it to promote myself,'[25] she said.

Physical and Social Transformation of the Urban Environment

Among all researched self-organised groups of citizens in Skopje, only one succeeded in achieving the goal and their requests were almost entirely fulfilled. In this case, the implications for the physical urban environment are evident and the transformation under economic reasoning failed. In addition, the group that fights against hunger and urban poverty, referred to above, gradually but persistently works on achieving its aim. However, they are realistic in how much can be achieved by single actions. Others are more critical in measuring the impact of their actions in the community, at least in its substantial form. 'It is realistic to say that some small "wins" have been achieved. However I am not sure whether many of the controversial actions that followed our "uprising" were somehow inspired by our protest,'[26] comments an interviewee.

24 Representative of 'Ploshtad Sloboda,' interview by Katerina Mojanchevska, February 27, 2014.

25 Representative of 'Spas za Vodno,' interview by Katerina Mojanchevska, April 04, 2014.

26 Representative of 'Ploshtad Sloboda.'

Yet, they all note that social effects on the community are more visible than the expected physical transformations. They agree that their engagement with the groups reinforced a sense of belonging to the community; a sense of responsibility to act for the good of the community; to turn distant neighbours into close friends; and to motivate the isolated into like-minded to other community members. Each of the interviewees stated that the gained experience contributed to their personal development, like a 'school of life,' something that is not taught in formal education.

The groups and their members are also an important model for the general public. Similar to the outcomes of Lima e Ferreira's Theatre of the Oppressed, they demonstrated that civic voice can make a difference in the community and that citizens cannot stand face to face with blunt injustices that harm their basic rights and needs. In some cases, the public showed solidarity with their initiatives and in a few, even institution representatives agreed with the cause of the civic groups but were powerless to initiate any changes.

In some cases, the action led to the appearance of negative social change. The absence of mass support for the community groups shows that people are hesitant and/or intimidated, passive and unaware and at times ready to 'exchange' the benefits of the community for their personal comfort. The inability of the groups to meet their goals accumulates a sense of helplessness and financial loss, making many people lose enthusiasm to be part of this civic activism. Some even thought to move from their community as a way to cope with the negative emotions accumulated by everyday acceptance of defeat. 'I cannot live with such injustice. Every day it reminds me that I am a victim and my house is devaluated,'[27] said one of the respondents. Some commented that this was precisely the goal of government: discouraging citizens to self-organise into groups.

Politically Sensitive Issues and Self-Organised Activism

There are no apolitical issues when it comes to urban planning of the city, because it directly affects the everyday life of citizens and invests their future and past into the present. However, some issues are more political than others or at least more controversial, as these related to the idea of new imagery of Skopje city centre.

Three of the groups were directly involved in the protest against changing of the city square and the destruction of green areas in the central area. All are remembered because of the direct confrontation with the institutions

27 Representative of 'Meksichka street,' interview by Katerina Mojanchevska, March 20, 2014.

including the police and judicial proceedings against members of the groups. These questions may have been local in terms that they were related to a particular part of the city (the main city square and the park around the hotel Bristol), but actually they talked about something much broader – the public space in general and how we deal with different opinions on about its appearance and purpose. The importance of this particular public space is that it is located in the city centre and as such belongs to all citizens of Skopje, not only to the immediate residents and nearby users. This fact contributed to a greater support, but also to a wide confrontation over the offered urban and visual solutions, as well as to the delocalisation of the issue.

Other groups, however, working on issues that were more of local significance, were not powerful enough to polarise the general public into 'supporters and counter-protesters.' Few of them have been successful in providing media support. However, this does not mean that because of the local nature neither of them was successful in achieving its goals. On the contrary, one of the groups fulfilled its mission, while another detected a case of corruption and abuse of power, which no institution is willing to resolve.

Funding of Self-Organised Initiatives

In general, the groups of citizens used self-financing as the main source of income. This is considered an important element of their independence. In few cases, some of the groups received small donations from an international donor and the contribution from within and outside the group was regarded as interference with the group's autonomy and neutrality. This influenced the groups' cohesion and their public image. Some groups noted that the absence of need for structural and continuous funding is exactly what distinguishes them from formal associations. However, in reality, the lack of funding creates difficulties for operation, especially when it comes to organising mass events, such as protests, public gathering of signatures or 'occupation' of public spaces. In general, the perception of 'spending' free time for unpaid civic activism is recognised as a major obstacle to social mobilisation.

Transformation of Goals

A critical question is what happens to groups that fail to fulfil their initial goal. Are they doomed to devolvement? An interesting transformation of the initial goals occurred in three of the self-organised groups. In the course of the activities their initial intention changed as the context and the power relations between citizens, institutions and the media changed. The change meant moving from specific to more general needs of a democratic society, such as the right to protest and the right to protect public property. Such transformation was an additional incentive for the groups' cohesion, enhanced the motivation and intensified the positive social impact on the community.

Media, Social Networks and Other Methods of Promotion and Support for the Group

In general, the groups had limited access to diverse mainstream media, either on local or national level. Each group understood the importance of media visibility and invested efforts in media communication. However, media 'gate-keeping' by political and economic elites caused their activities to seem mostly invisible.

Social networks, especially 'Facebook' were largely used to facilitate mobilisation and public support. Most of the groups emphasised the importance of the new media that helped them to inform and mobilise the citizenry. Also, establishing 'hotlines,' organising concerts, exhibitions and the like, increased their visibility and disseminated information to the public. At the same time, the groups were aware of the difference between real-time and 'Facebook' support.

The Role of Political Elites in Jeopardising the Autonomy of Self-Organised Groups of Citizens

Self-organised groups of citizens have conflicting views on whether the group should communicate and cooperate with political parties to achieve their goals, how to deal with the political affiliation of the members of the group and whether the political commitment compromises their positions and impartiality.

At the same time, connecting to a political party is a stigma and fear for the self-organised groups. In the highly politicised context of Macedonia, this usually refers to connecting with opposition political parties who, on one hand, are considered democratic, open-minded and supportive to civic activism but on the other hand as hijackers of civic activism and abusers of the honest civil interventions. Finally, there is antipathy between groups equating civil with political activism as in the present context political activism is solely associated with membership to a political party. Some groups even feel resentment to be associated with any political party and feel that those groups that do so are partly responsible for the perception of civic activism as politically 'tainted.'

It seems that cooperation with political parties may be unfavourable and tricky, but above all, carefully planned and guided. At the same time, it should not benefit only a few in the group but be more in the spirit of a communal interest.

Barriers in Structuring and Operation of Self-Organised Initiatives

There are barriers that arise from the form of self-organisation, which affects their structure and activity. 'Fighting a battle is a full-time job,'[28] said one respondent. With no external funding and negotiating between paid and unpaid responsibilities, respondents agree that it is difficult to maintain the group cohesion, feel motivated, react in a timely manner and propose feasible solutions. If the complex media environment decreased social and economic power of the citizens and the threats of ethnic polarisation (which in truth during the interviews was not mentioned by any of the respondents) are considered, the conditions allowing self-organised groups of citizens to be active and to be heard are both difficult and challenging.

Unintended Effects of the Self-Organised Activism

The phenomenon of organising a 'counter-protest' in the political culture in the country can be associated to the civic activism of the self-organised groups. Counter-protests developed as a governmental response and attempted to deal with civic actions that contest or oppose the government's decisions. Apparently the political elites understand the potential of the self-organised groups of citizens to motivate and aggregate civic discontent. One interviewee noted: 'the tradition of organising counter-protests began in March 2009 along [with] the first architectural uprising. Afterwards, the same method was used to abort various civic dissatisfactions obstructing the constitutionally guaranteed right to protest. As a result, activists and especially informal groups distanced themselves from the position of being political actors. This is detrimental! It demonstrates who has the right to an opinion and can organise the citizens. The perception that it is impossible for citizens to self-organise must change.'[29]

Networking and Follow-Up

Through their activities, all self-organised civic groups have shared their experiences with numerous organisations, platforms and forums. There were efforts in establishing close collaborations between like-minded groups. Unfortunately, despite having common activities, the groups failed to capitalise and amplify the achievement of their goals. The groups' activities have been largely perceived as local with a strong connection with their micro community, which in a way prevented them from acting more widely and sharing responsibility for the problems of those outside of their immediate environment.

---

28 Representative of 'Ploshtad Sloboda.'

29 Representative of PAB, interview by Katerina Mojanchevska, March 18, 2014.

Basically, all these attempts are unfinished efforts to create a more integrated civic platform, which can advocate for new models of collaboration between citizens and authorities.

Due to various influences, the groups have reduced in number, disappeared or changed their mission and legal status. Some are attached to other forms of professional or civic associations. Others are more active as individuals than as a group but are engaged in the transfer of knowledge and lessons learned. Nonetheless, they remain persistent in the idea to be 'a thorn in the government's side.'

The respondents agree that they operate in complicated political, economic, social and cultural context, yet, they continue to believe that institutional re-design that promotes citizens' participation in creating policies will eventually affect the decision-making process.

6 Conclusion

Today, as is also apparent in the other chapters in this section of the volume, in order to understand the nature of urban space it is not enough to make reference to the architectural tradition and normative approach to the built environment. The growing importance of the public space as a 'combination of people, buildings and events'[30] puts forward the citizens' involvement as a critical factor in imagining the city building practices.

The production of space and Henry Lefebvre's writings remind us that city space is more than an 'empty container,' it is a system of socially produced meanings which derive from knowledge, social and cultural relations and everyday life experiences – all which are subject to socially constructed processes.[31] Space is both shaped by the dominant interpretation of knowledge and the social and spatial relations and experiences. At the same time, the involvement of these elements in processes of identity formation, shape and influence power struggles among dominant and excluded groups. Furthermore, Lefebvre is important for embracing diversity arguing for a differential space as opposed to the abstract space and for the 'right to be different'[32] along the 'right to the city.' A point also made by Smith's and Steinø's disruptive, 'other'

30 Ali Madanipour, *Designing of Urban Space: An Inquiry into a Socio-Spatial Process* (West Sussex: Jonh Wiley and Sons, 1996), 3.

31 Madanipour, *Designing of Urban Space*, 19.

32 Madanipour, *Designing of Urban Space*, 19.

uses and creation of spaces of exchange and education[33] and Lima e Ferreira's empowering performances that transform the oppressed into protagonists.[34]

The research showed that public space in Skopje is produced predominantly through the views of abstract space makers and in a process that represents the power and ideology of the dominating groups. In a government-led process of participation, the voices, experiences and visibility of citizens is undermined and frequently used to legitimise planners' positions and ideas. In such conflict, the public space relegated to be of equal access and public involvement, in fact does not create genuine institutional arrangements and practices that stimulate citizens' participation in city-making and does not create opportunities to exercise real control over city planning and management processes.

Critical themes that motivate citizens' interest are those of destruction of the immediate built environment, destruction of the social functions of the public space, privatisation of public goods and deteriorated quality of life. However, wider platforms of citizen dissatisfaction are difficult to organise and sustain. Political pressure, institutional obstructions, delegitimisation, incorrect media representation and fluctuation of personal and group motivation are considered among the main obstacles of the local self-organising micro networks in the urban and social planning of the neighbourhoods. Their powers lie mainly in mobilising the shared sense of community, responsibility and building of the social and cultural capital of the neighbourhood. Some social groups are perceived more powerful when self-organised and this is mainly due to their personal capacities, social responsibility and political awareness. One difficult issue still remains: how to reach disadvantaged groups and their 'right to the city.' The 'Theatre of Oppressed' method vividly elaborated in this volume offers a practice of bringing closer less-privileged citizens and city-making using popular education for emancipation and practice of citizenship.[35]

The power-distribution between official planning and self-organised networks for urban development lacks transparent communication and balanced power division among planners and users and inhabitants of the space (e.g. the diverse ethnic groups). The conceived vision by planners and politicians of what constitutes 'appropriate' plans of public space may and, should be, allowed to be contested by the sensory experiences in public space and the actual lived practice of place by those using it.[36] These two aspects of space, that of the planners and the other of the users, may be also in conflict as

33 Smith and Steinø, in this volume.

34 Lima e Ferreira, in this volume.

35 Lima e Ferreira, in this volume.

36 Don Mitchell, 'The End of Public Space? People's Park, Definitions of the Public, and Democracy,' *Annals of the Association of American Geographers*, 85.1. (1995): 115.

planners' ideas, governmental suggestions and commercialised intentions may differ from citizens' needs and how they think about the public space, its function and use. In such perspective, public space planning process should have the capacity to allow interpretation and reinterpretation of conceived regulations through joint, inter-dependent and shared activities.

The research results confirmed the important role of citizens in creation of local policies and solutions, which are flexible to the diverse needs of the communities in the neighbourhoods and the city. Citizens and activists consider that they can make change in the local setting by participating in the institutional architecture of decision-making. 'Anyone can be a city-maker' comment Topaloğlu and Çebi[37] in this volume. The enduring disappointment with technocratic planning and empowerment of citizens supports the belief that the civil society and self-organised citizens' groups can be the key to social transformation.

However, the power of self-organisation is strongly contested. One of the main criticisms is the influence of political parties and business interest, which question the impartial assessment of social reality. It seems that this criticism is relatively easily transferred in the perception of self-organised groups of citizens. Negatively labelled, such forms of organisation struggle with internal and external contestations and deriving from the diversity and variety present in the group, fall short in mobilising wider social support. A genuine question remains: Can self-organisation be an answer to these deficiencies of the civic activism in the country?

From a different perspective, citizens involved in activism and self-organisation prove that there is a need for a new relation between the citizen(s) and the institutions. They show that it is possible to democratise the public sphere from below and include diverse segments of society but only in situations where there is formal political will to do so. Self-organisation bridges the gap between formal and informal urban development and is capable of (re)vitalising or giving new meanings to urban spaces. They manage to narrow down the gap between the legal and technical side to institutionalise participation and the reality of effective involvement, as noted by Cornwall and Coelho.[38] Yet, this is only evident in situations where there is a dialogue between these two sides and political determination.

The wider community, the media and everyday politics exert immense influence over the group dynamics and effects. The political decisions in how to manage the relations with these sectors, determines the extent and

37 Topaloğlu and Çebi, in this volume.

38 Cornwall and Coelho, 'Spaces for Change?' 3.

maintenances of achieved results. Self-organisation should serve the civic interests of its constituencies and it should employ transparency and wider information sharing of the goals, activities, structure and strategy. In this quest, the role of the social media should not be under-estimated but rather fully used in developing information and knowledge sharing platforms.

Therefore, it is fair to say that the realisation of the full potential of self-organisation is yet to be exploited. A dilemma remains to be further researched. Despite the importance of citizens' participation in local urban processes, civic engagement is declining. Factors such as lack of time, misbalance between professional and free and family time, accumulated with the sense of fear and political revenge, impede social responsibility towards community improvement.

Acknowledgement

The research has been conducted as part of the project 'The Initiator, the Artist, his Advocate and the Urbanist (IAAU) – a platform for culture and city-making,' implemented by Coalition for Sustainable Development – CSD, Macedonia; Co-Plan, Albania; School of Urban practices, Serbia; URBEGO, Denmark and Block 74, Netherlands. It was mentored by Ms Sonja Damchevska (CSD).

This chapter is based on a chapter included in the following text: Mojanchevska, K. (2017) 'Urban planning and self-organized citizens' networks in post-transitional societies in south-eastern Europe: a case study of the city of Skopje,' in: Van Dijk, M.P., Edelenbos, J. and Van Rooijen, K. (eds). *Urban Governance in the Realm of Complexity* (https://doi.org/10.3362/9781780449685).

Bibliography

Banerjee, Tridib. 'The Future of Public Space.' *Journal of the American Planning Association* 67.1 (2001): 9–24.

Booher, David E. and Judith, E Innes. 'Network Power in Collaborative Planning.' *Journal of Planning Education and Research* 21.3 (2001): 221–236.

Boonstra, Beitske and Luuk Boelens. 'Self-organization in urban development: towards a new perspective on spatial planning.' *Urban Research & Practice* 4.2 (2011): 99–122.

Cornwall, Andrea and Vera Schattan P. Coelho. 'Spaces for Change? The Politics of Participation in New Democratic Arenas.' *Spaces for Change? The Politics of Participation in New Democratic* Arena, edited by Andrea Cornwall and Vera Schattan P. Coelho, 1–29. London: Zed Books, 2007.

Creswell, John W. *Research design: Qualitative, quantitative, and mixed methods approaches*. Thousand Oaks: Sage, 2009.

Cuthbert, Alexander R. *The Form of Cities: Political Economy and Urban Design*. Oxford: Blackwell Publishing, 2006.

Doucet, Isabelle. 'Negotiating Brussels: collectives in search for a common world?.' *Urban ontologies: importing ANT into the urban studies seminar*. Berlin, Germany: Centre for Metropolitan Studies, 2007.

Fainstein, Susan. *The Just City*. New York: Cornell University Press, 2010.

Fraser, Nancy. 'Rethinking the Public Sphere: A Contribution to the Critique of Actually Existing Democracy.' *Habermas and the Public Sphere*, edited by Craig Calhoun, 109–142. Cambridge: MIT Press, 1992.

Healey, Patsy. *Collaborative Planning: Shaping Places in Fragmented Societies*. Basingstoke: Macmillan, 1997.

Innes, Judith E., and David E. Booher. 'Collaborative Policymaking: Governance Through Dialogue.' *Deliberative Policy Analysis: Understanding Governance in the Network Society*, edited by Maarten Hajer and Hendrik Wagenaar, 33–59. Cambridge: Cambridge University Press, 2003.

Klekovski, S., Buldioski, G., Bitikji, F., Nuredinoska, E., Gaber-Damjanovska, N., Zajazi, K., Sazdovska, S., Gerasimovska-Kitanovska, B. (2006) *Po 15 godini tranzicija – Od stabilizacija kon gragjansko uchestvo* (After 15 years of Transition – From Stability to Citizen's Participation). Skopje: Macedonian Center for International Cooperation.

Madanipour, Ali. *Designing of Urban Space: An Inquiry into a Socio-Spatial Process*. West Sussex: Jonh Wiley and Sons, 1996.

Mitchell, Don. 'The End of Public Space? People's Park, Definitions of the Public, and Democracy.' *Annals of the Association of American Geographers* 85.1 (1995): 108–133.

Pløger, John. 'Strife – urban planning and agonism.' *Planning Theory* 3.1 (2004): 71–92.

Sandercock, Leonie. 'When Strangers Become Neighbours: Managing Cities of Difference.' *Planning Theory and Practice* 1.1 (2001): 13–30.

State Statistical Office of Republic of Macedonia. *Use of ICT in households and by individuals* (Koristenje na informatichko-komunikaciski tehnologii vo domakjinstvata i kaj poedinci). No: 8.1.14.28, October 2014. Available at: http://www.stat.gov.mk/pdf/2014/8.1.14.28.pdf [20 March 2015]

Watson, Vanessa. 'The Usefulness of Normative Planning Theories in the Context of sub-Saharan Africa.' *Planning Theory* 1.1 (2002): 27–52.

PART 3

Temporariness: Artistic Interventions on Space and the Process of Becoming

CHAPTER 9

Perpetual Transitions: A Communication between Photography and Absence within the Spatial Experience of the Ruin

Clementine Monro

Abstract

Perpetual Transitions is a practice-based research project that investigates how absence is observed through the spatial experience of the architectural ruin and the communication of that experience through the medium of photography. It is addressed through an experiential study and photographic response to the Georgian ruin of Nettleham Hall, England. The word *ruin* has its origins from the notion of 'falling or fallen stones.'[1] In this context, ruins are seen as the fragmented remains of man-made architecture. In experiencing the ruin of Nettleham Hall within its present state, absence has the capacity to embody lived space and ephemerality of the ruin. The photographic response as an experiential inquiry is key to the communication of absence and the phenomenologically experienced ruin. The capacity of photography to communicate time, duration and the ephemeral mediates a visual trace and expression of the felt experience of absence. My questions of how is absence felt and experienced within the ruin, and how does one communicate absence through photography, seek to connect and then transfer the felt experience of the ruin into an art piece that addresses absence and spatial experience.

1 Introduction

This research focuses on how absence is observed and registered within a specific and phenomenological setting, the architectural ruin. In order to respond to the research question as a whole, it is broken down into two parts: *How is absence felt and experienced within the ruin,* followed by *How does one communicate absence through photography.* This deliberate action has allowed for a more rigorous investigation and contemplation into the areas of

1 Michael Roth, Claire Lyons, and Charles Mereweather, *Irresistible Decay: Ruins Reclaimed* (Los Angeles: Getty Research Institute Publication, 1997), 1.

 | DOI 10.1163/9789004365520_011

phenomenology and ontology that the primary research of felt experience has taken while underpinning the significance and meaning of the practice. This research has looked at absence within a specific and phenomenological setting, the architectural ruin of Nettleham Hall. The first part of this chapter positions and foregrounds the ruin as a place of fascination and the rationale of basing the research and practice in this specific setting, before addressing the meaning and nature of Nettleham Hall as a ruin, locating and contextualising it as a specific place and space for the experience of absence.

2 The Makings of the Ruin

When a building or structure ceases to function as it was originally intended, and, if left alone, it starts a slow process of decline and decay. Over time the building may become structurally unsound, where nature is allowed to invade seeking out exposed spaces and small cracks in which to take root and grow. Over an extended period of time the building is propelled into a further state of decay, where established nature alters and reshapes the building by splitting, shifting and subsequently pulling down structural work with overgrown and heavy vines. The building can be further altered through interventions. such as extended exposure to time, animals, human activity and weather. At this point, where decay turns the building into a ruinous state, the building becomes a ruin.

3 The Ruin: A Place of Fascination

It is the intention that in observing the current behaviour of fascination towards the ruin, the contextual elements of phenomenological inquiry that define this research, absence, spatial experience and ruin are more clearly defined. It positions the ruin at the forefront through which absence and spatial experience unfold and are recorded by multiple methods of phenomenological inquiry. The methods of research, observation, photography, theoretical analysis and reflection, serve to offer mediation between the space of Nettleham Hall and the researcher facilitating a dialogue of experience and practice taking place.

There are two types of ruin that are worth considering for the placement of where this research lies: the *evolving* and the *arrested* ruin. Each differs in both appearance and experience. The ruin that is in an evolving state of decay, left to the devices of nature, is allowed to follow the course of time and slowness of entropy. In a constant state of flux and perpetual transition, they are positioned in the present, the here and now. Philosopher Dylan Trigg echoes this notion of present time by suggesting, 'in a word, they *decay.* The result of this active decay is that they viscerally engage with our senses.'[2] This unfolding relationship with time and therefore richer awareness of existence, mediates the experience of absence as real, a continued authentic state of becoming. It is not possible with the arrested ruin because it has been preserved from further decay as soon as it has been reclaimed by the heritage industry, such as English Heritage, formed in 1983 and The National Trust, formed since 1895, in England. The type of arrested ruins in question are the heritage monuments, buildings and structures that although attract millions of visitors from around the world, are 'mere novelties and outside of time.'[3]

Through the course of this research I have visited, experienced and photographed both types of ruin. For me, there is a marked difference in ontological experience, spatial awareness and sense of absence. In the arrested ruin, I have not felt the same experiential connection that I felt with the evolving ruin. My own fascination and rationale for visiting the evolving ruin of Nettleham Hall is rooted in the phenomenological experience of its present state. It is a very real experience that fills the senses with a rich awareness of existence and time as unfolding. This ontological experience is where the rationale lies for basing the research and practice within the evolving ruin and at Nettleham Hall in particular.

4 Absence: A Felt Experience in the Ruin

We experience and perceive absence through ourselves, our consciousness and, in particular, awareness of the senses. The perception of architectural space involves the body as an entirety, as Philosopher Merleau-Ponty suggests in *The Phenomenology of Perception.* Paralleling Katsuhiko Muramoto's[4] theoretical assertions in this volume, when we perceive space and place all our

2 Dylan Trigg, 'Architecture and Nostalgia in the Age of Ruin,' (Viewed 7 December 2014). https://www.academia.edu/208447/Architecture_and_Nostalgia_in_the_Age_of_Ruin, 2.

3 Dylan Trigg, *The Aesthetics of Decay: Nothingness, Nostalgia and the Absence of Reason,* (Peter Lang, 2006), 183.

4 Katsuhiko Muramoto, 'In Search of a Relational Body: Reflections on Sustainable Architecture,' in this volume.

senses are affected.[5] Through awareness of the senses our bodies react physically and emotionally to the ruin, this reaction is felt through mediation of body with that particular place. In this context of *ruin*, place is referred to as 'individuated and particular'[6] and 'unlike space, which is geometrically defined and homogenous, place is local.'[7] Being situated in place within the ruin, absence holds form, meaning and significance. Absence is described not as a tangible thing or object but as a feeling or felt sense. It is a manifestation of feeling shaped through the body's awareness, of the felt experience within the present state of the ruin. Absence is both experiential and phenomenological, meaning we can observe absence through direct experience of place and space. By giving absence substance it has the capacity to embody the fullness of presence.

Like absence, place is not a thing or actual object that exists in the world, it is not something that we go out and locate. Place exists because our bodies interact with it and make it so. In a subsequent chapter in this volume Anthi Kosma alludes to the body 'as a form by itself, the body as a generator of space.'[8] In this context, place relies on our bodies to give it substance and meaning. This is furthered by philosopher and critic Dylan Trigg: 'we live in and remember places, because the particularity of place defines our temporal dwelling, and so our continuous identity.'[9] Merleau-Ponty also suggests that 'place is the means whereby the position of things becomes possible.'[10] It is through the body's experience and interaction with being in a particular given place that makes this feasible. He describes the body as 'the vehicle of being in the world, and having a body is, for a living creature, to be intervolved in a definite environment.'[11] The felt sense of absence is possible through the body's reaction and interaction to the space that is particular to the ruin. This often gives us nothing of artefacts and facts it is, instead, an often, seemingly empty space yet through absence and the embodiment of presence it is full. The immeasurable sense of

5 Maurice Merleau Ponty, *The perception of Phenomenology*, trans. Colin smith (London: Routledge, 2006), 282.

6 Dylan Trigg, *The Aesthetics of Decay: Nothingness, Nostalgia and the Absence of Reason*, (New York: Peter Lang, 2006), 121.

7 Ibid. 121.

8 Anthi Kosma, 'Spacing by Trace: Four Events of an Out of Place Experience,' in this volume.

9 Dylan Trigg, *The Aesthetics of Decay: Nothingness, Nostalgia and the Absence of Reason*, (New York: Peter Lang, 2006), 121.

10 Maurice Merleau Ponty: *The perception of Phenomenology*, trans. Colin smith (London: Routledge, 2006), 282.

11 Merleau Ponty: *The perception of Phenomenology*, 96.

absence within a ruin can also be felt through the presence of the remaining space. An example of this could be the body's keen awareness in detecting the way jagged natural light moves through curious open crevices and surfaces within the building's remaining structure capturing Trigg's notion of temporality: 'We are forever in the here, and it is from that our experiences take place,'[12] the here being the present.

5 Spatial Experience

The experience of visiting ruins can be read through Trigg's phenomenological perspective, an experience felt through the body and the body's senses:

> Our bodies respond to the texture of ruins. It is a very tactile and visual experience; from the unevenness of the floor beneath ones feet to the presence that a ruin possesses.[13]

It is through the body that multiple sensory information such as vision, touch, hearing, smell and taste interact with the environment around us creating a viscerally emotional response to react and interact with the world. Senses such as vision, sound and touch are essential for spatial experience. Responding to the spatial experience of the ruin through the senses stimulates the body's senses and correspondingly the ruin also stimulates the imagination. As Maria Joáo Duráo suggests: 'Between body and place there is much more than position, there is the data of our experience and expectations.'[14] Fascination with ruins has led to multiple phenomenological interpretations through art, poetry, photography and sustainable architecture. Critic Brian Dillon suggests that the reason lies in the ruins 'radical potential – its fragmentary, unfinished nature is an invitation to fulfil the as yet unexplored temporality that it contains.'[15] The architectural ruin, a vehicle for spatio-corporeal experience,

12 Dylan Trigg, *The Memory of Place: A Phenomenology of the Uncanny,* (Ohio University Press, 2012), 4.

13 Dylan Trigg, 'Architecture and Nostalgia in the Age of Ruin,'(Viewed 7 December 2014) https://www.academia.edu/208447/Architecture_and_Nostalgia_in_the_Age_of_Ruin, 2.

14 Maria Joáo Duráo, *Embodied Space: A Sensorial Approach to Spatial Experience,* Propulsion & Energy Sciences International Forum: SPESIF-2009. AIP Conference Proceedings, Volume 1103. AIP Conference Proceedings, Volume 1103, Issue 1, pp. 399–406. (Viewed 15 January 2015) http://www.astrosociology.org/Library/PDF/Durao_SPESIF2009.pdf.

15 Brian Dillon, introduction to, *The Ruin: Documents of Contemporary Art,* edited by Brian Dillon (Cambridge: MIT Press, 2011), 18.

offers an observable tactile environment for the investigation of the ontological significance of absence.

In order to explore the potential of these ideas of spatial experience and meaning of connectedness, the final part of this section serves to situate the remains of Nettleham Hall as a ruin and rationalise it as a place and space of importance for the experience of absence and photographic communication. This place was once a home, intact and lived in and that is what I first connected with and subsequently why I decided to focus on for the research. Subsequent visits to the Lincoln Archive Centre provided photographs of the interior and exterior of the property whilst it was inhabited and provides evidence that it was home. As poignant as it was to be able to view the fragile photographs of the late nineteenth and early twentieth century, as it had not been possible to until recently, it has not changed or altered the experience that I have with the current state of the ruin. Although I hold a fascination and appreciation of the history and meaning of Nettleham Hall, it is in its current and evolving state as a ruin that an ontological examination of a sense of absence and spatial awareness can occur.

Nettleham Hall will not remain standing in the long term and every visit sees more of the structure in heaps of limestone and brick rubble on the woodland floor. There is no future or renewal for this ruin. It is testament to the decline of so many buildings of the same era, where many properties became too big to maintain, when families moved to the cities or the rising housing tax on large properties forced families out. In documenting this research and in creating an art installation as a response, the essence of Nettleham Hall may well remain standing for a while longer.

6 Photography Practice: A Phenomenological Inquiry into the Investigation of Absence

The ontological nature of photography and the phenomenology of it can be seen as a methodological approach to the encounter of absence as experienced in the ruin. Photography has the capacity to communicate and visualise a given object or situation but this research goes further than viewing the photographic image as a given and places it firmly within phenomenology and spatial experience. It also looks at how photography acts and reacts as a phenomenological and methodological approach to how absence emerges in the ruin and how this emergence is subsequently communicated into a visual form. In *Imprinted on the Body: Changing Sensibilities in Time and Spaces,* Shona Hill follows Toffoletti's work of new technologies as a reformation and transformation of 'earlier technologies, such as photography' that have 'influenced the way we have come to understand ourselves.'[16] This reformation of photography has seen it transform from a traditional document of observation to an extension of body experience and phenomenological medium of communication. By situating the body at the centre of how we respond to the experience of absence the medium of photography can then visually document and thus communicate that experience. Photography has the capacity to contribute a heightened sense of experiential quality of the ruin through capturing and recording duration and time within a given space. Similarly, in this volume Kosma describes the 'experience of drawing as an interior matter' that it is an, 'experience that disappears when the action stops, a performing art.'[17] In her seminal text, *On Photography,* Susan Sontag suggests photography's phenomenological ability to capture experience and that the camera acts as an extension of the body.[18] The photographic response as an experiential inquiry is key to the communication of absence and the phenomenologically experienced ruin of Nettleham Hall in this research. It is the primary reason for choosing photography as an essential method of phenomenology and visual response to the ruin. It allows photography to act as an experiential element of this research from the perspective of making, into a corresponding and responsive art installation.

In its widest sense, the photograph is a visual trace, a communicator of time and duration. It is a combination of light and duration of exposure to time that determines a resulting image. That being said, my emphasis with photography is not motivated by the technical capabilities of the camera but from a more existential concern. In the text *Locating Memory: Photographic Acts,* Kuhn and McAllister allude to photography as seen by Walter Benjamin and Siegfried Kraucer as

> … a visual technology, that raises existential and epistemological questions-questions about the nature of existence and about how we can know the world, questions about how photography can become implicated in our knowing the world.[19]

16 Shona Hill, 'Imprinted on the Body: Changing Sensibilities in Time and Spaces,' paper presented at the '4th Global Conference on Time, Space and the Body.'

17 Kosma, in this volume.

18 John Berger, *About Looking,* (London: Bloomsbury, 1980), 53.

19 Annette Kuhn and Kirsten Emiko McAllister. *Locating Memory: Photographic Acts.* (Remapping Cultural History, Volume 4. New York: Berghahn Books, 2006), 5.

As early as the beginning of the twentieth century, the photographer Eugene Atget made a series of photographic records of Paris. He 'refused to draw on established structures of seeing – bourgeois genres and aesthetic forms – and instead sought to generate new modes of perception.'[20] This can be seen as a phenomenological approach communicated through photography and echoes what Trigg says about *being in the world*, in that, 'we are forever in the here, and it is from that, that our experiences take place.'[21] Photography can act as a communicator in how we know and interact with the world. The images of Paris by Atget are dense with experience and offer a visual response and interpretation of both past and present.

Photography is also about duration not in terms of the mechanical process of photography but what duration alludes to. Traditionally the photograph is taken with a relatively short duration, such as Japanese photographer Hiroshi Sugimoto's static depiction of the *Tyrrhenian Sea* in the *Seascapes* series. On the other hand, duration might be extended resulting in a blurred image from the same series. Both examples show the camera's mechanical time of duration but more significantly, both examples also communicate duration of time. This communication of time is what raises the complex question of the relationship between time and the photograph and thus highlights the fundamental nature of photography. The film theorist, Peter Wollen describes duration within a photograph as follows:

> Images, may themselves appear punctual, virtually without duration, this does not mean that the situations that they represent lack any quality of duration or other qualities related to time.[22]

Sugimoto also sees the photograph as an object in a state of becoming and crucially, in an experiential state. Sugimoto's large-scale photographs of *Movie Theatres* address persistence of time as the photographs are achieved through long exposures. These exposures of cinema screens convey not only the time of duration, the photographic exposure would last the length of the feature film. Sugimoto more poignantly conveys duration of time within a single image. At first glance the image appears to be a traditional photograph of short duration, a still and clear image of an old style movie theatre with empty seats. However, this apparent stillness is juxtaposed by the blindingly, bright white

20 Kuhn and McAllister. *Locating Memory: Photographic Acts*, 3.

21 Dylan Trigg, *The Memory of Place: A Phenomenology of the Uncanny*. (Ohio: University Press, 2012), 4.

22 Peter Wollon, *Fire and Ice*, (in Photographies, no 4. Paris: April 1984), 109.

light of the cinema screen, which is only possible because of the long exposure. These photographs contrast but only in a visual sense the static *Seascapes*. Most of this series consists of long exposures, typical of Sugimoto's practice, where the *Seascapes* are blurred to the point where there is no defined horizon between the sea and sky. Both the series *Movie Theatres* and *Seascapes* evoke a sense of time that is a careful combination of the production of the image and the reception of the image regardless of the mechanical time of duration. This provokes an understanding of time as a state of becoming, a fluid event that challenges the notion of photography as a static moment of the past. These photographs raise the complex question of the relationship between time and the photograph, which goes beyond temporality and thus highlights the fundamental nature of photography as a phenomenological expression of a given experience. The *Seascape* series by Sugimoto are a good example of spatial experience captured and framed into a single photograph.

An example of the ontological nature of photography and notions of time and ruin within a photograph are the images by Robert Polidori of Pripyat and Chernobyl. A photographer of chiefly human habitats and environment, Polidori produces large-scale photographs that capture and tell of human experience. In 2006 Polidori documented what was left after one of the nuclear reactors at Chernobyl exploded in 1986. The rooms photographed at Chernobyl and nearby town of Pripyat are shown as a metaphor of lives past and traces of people in a space that reveals the texture of absence. Captured by the camera, the images invite the spectator to experience the feeling of being in that space.

Photography, in its widest sense, encompasses both still and moving image. Within my own practice for this research I experimented with different photographic equipment and began filming at the hall. As such, it is important to point out that filmmaking is an extension of photography and acts in a similar context of visualising absence and experience. Filmmaker and critic, Peter Wollen, alludes to film making as a visualiser of the *here and now*, where he describes 'photography as a point, film like a line.'[23] Film can be described as a sequence of movement therefore it becomes more synonymous to the notion of becoming. Siegfried Kraucer's well-known book, *The Theory of Film,* suggests that film is representative of reality as it evolves in time by use of separate images in a sequence.[24] In other words, in a cinematic film, time can be looked at simultaneously as past and present. As well as the still photograph, film is another way that communicates experience as it occurs. Tacita Dean's film piece,

23 Peter Wollon, *Fire and Ice*, (in Photographies, no 4. Paris: April 1984), 108.

24 Siegfried Kraucer, *The Theory of Film: The Redemption of Physical Reality*. (Oxford: University Press, 1960), 41.

Kodak 2006, offers an effective use and treatment of photographic filmmaking that is significant to translating the spatial experience of a defined space. The film also touches on the notion of ruin and how one can communicate a phenomenological experience and successfully translate that experience into a film piece. Dean's choice of analogue over digital filmmaking is the fact that she chooses to use the material of film, which is now something of the past. Dean's work speaks of nostalgia for technologies and artistic forms of the past as practices about to become obsolete. On discovering that the Kodak Corporation was about to cease production of black and white standard 16 mm film Dean purchased the last of the stock and filmed, using both black and white and colour film, the production process at the Kodak factory in Chalon-sur-Saone, in France.[25] The result is a forty-four minute narrative that follows the making of celluloid film as it runs through several miles of machinery. The film piece reflects on the fading medium of analogue film itself. What is important about this film piece and significant to how photography can communicate absence and phenomenological experience is the way in which Dean creates tension and movement within the film piece without any sort of action.

When we think of a film we might think of it as containing an establishing shot having a beginning, middle and an end. Dean's film *Kodak 2006* does not contain any of these common features associated to film, it is simply a day in the life of the factory. Rather than telling a story, Dean is showing a narrative that is slow to unravel. The first thirty-five minutes are concentrated on the production of celluloid film and is aesthetic in its approach. Much of this section of the film includes close up shots that are both abstract and monotonous. There is a real sense of duration and persistence of a technology through repetition of image sequence. Throughout most of the film, the narrative is allowed to slowly unfold and reveals glimpses of human trace and absence. In a subsequent chapter in this volume, Kosma refers to the act of drawing as a 'corporeal experience'[26] of human trace. The last ten minutes of Dean's film encompasses a very different sense of the spatial experience of the factory in a negative way. This is perhaps testament to the decline of the factory and an industry that is coming into ruin. The rest of the film piece shows quick successions of sequence of broken furniture and rubbish scattered across much of the abandoned film packaging section. The lighting is harsh and evokes a rather angry aesthetic and spatial experience to the first portion of the film, which is composed in its monotony and repetition. By allowing the narrative

25 Brian Dillon, *Ruin Lust*, (Catalogue for the exhibition *Ruin Lust*, Tate Britain, 4 March–18 May, London: Tate Publishing, 2014), 52.

26 Kosma, in this volume.

to develop in this way, the significance of absence, temporality and becoming are highlighted.

Similar, to the narrative of Dean's *Kodak, 2006*, are the photographs of Jane and Louise Wilson. Their phenomenologically based photography both responds and reacts to uninhabitable and abandoned places. Through the communication of still photographs and film footage the places in question, whether the remains of the Northern Ukrainian city, Pripyat, or *Stasi City* the headquarters of the former East-German secret police, the art works created by Jane and Louise Wilson provide the viewer a way to inhabit and experience the uninhabitable. This is made possible through large-scale multi screened art installations, which highlight the uninhabitable as symbolic spaces of limbo through events, such as radiation, abandonment and subsequent decay. In the photographs of Pripyat, part of a series from: *Atomgrad: Nature Abhors a Vacuum* series and similar in one sense to Polidari's work, is that the remains of the town are exposed and 'laid bare.'[27] Whereas Polidari focused his work around the subject of human trace and what is left behind at the nuclear site of *Chernobyl*, the photographs of neighbouring town *Pripyat* made by Jane and Louise Wilson offer a further vision. Their photographs focus on public spaces, such as the hotel, cultural palace, library and swimming pool of the town and situate a yardstick in each shot. These places were part of the housing spaces for the workers of Chernobyl and seen as a contemporary community of the future. By placing a yardstick in each image, the photographers also place the body into that space. The yardstick is also representative as a measuring scale of radioactivity and the passage of time.[28]

The film piece by Tacita Dean and photographs by Jane and Louse Wilson, allude to the ontological nature of photography as a phenomenological response in its making, visualising and subsequent viewing. The series, *Seascapes* and *Movie Theatres* by Hiroshi Sugimoto, challenges the traditional idea that the photograph is a static moment, such as Barthes theory of the *Punctum*, where the photograph is seen as a temporal hallucination, a product of something that 'is not there' but 'it has indeed been.'[29] Sugimoto sees the photograph as an event evolving or time as becoming. It is also the signifier of phenomenological experience, similar to Polidari's large-scale photographs, which also

27 Guiliana Bruno, *Modernist Ruins, Filmic Archaeologies*, 2003. Lost Sites of Power, Ruins: A Document of Contemporary Art, edited by Brian Dillon (Cambridge: MIT Press 2011), 77.

28 DCA Dundee, 'interview with Jane and Louise Wilson,' YouTube. Online video clip, (viewed February 27 2015) http://www.youtube.com/watch?v=Wm4vl6DhUXY.

29 Roland Barthes, *Camera Lucida*. (London: Vintage, 2000). First translated by Hill & Wang, 1981, 115.

signify and embody spatial experience. This further alludes to the photograph as a phenomenological sense of being in the here and now and thus embodying a lived and spatial experience.

7 Choosing the Ruin

As part of the initial investigative process I visited a number of ruins including industrial, urban, rural and arrested. To form a convincing rationale of choosing one site where the practice and experience of absence would take place, it felt necessary to investigate a variety of spaces beforehand. The option of basing myself at a number of ruined spaces would have produced interesting but fragmented results and would not have provided a suitable nor cohesive environment for the corresponding art installation that I had in mind.

Early in the process of observing and photographing different sites, the significance of the dissemination was informing the way I photographed spaces as I was experiencing them. This level of inter-connectedness between practice and installation became routine. As I became aware of and experienced absence I found I was able to remove myself from just making objective decisions, of how to operate the mechanical practice of photography and instead, allowed space, absence and experience inform the practice and process of making. In the book *About Looking,* John Berger refers to photography as something that is capable of achieving the 'Now' if it is created in the right context.[30]

Reflecting on Berger's assertions in relation to the various sites that I experienced and photographed at the start of this research reinforced the need to find a suitable location where I felt connected and was able to access. I have visited the ruin of Nettleham Hall countless times over the past six years and am familiar with the place and space. It is set within a secluded rural setting and is readily accessible. I felt a connection to this place, which I was not fully aware of at the start but as I observed the space over time the connection evolved into a sense of place and *home.* In his seminal text *The Poetics of Space,* Gaston Bachelard refers to *home* as a shelter and suggests that

> The House, quite obviously, is a privileged entity for a phenomenological study of the intimate values of inside space, provided, of course, that we

30 John Berger, *About Looking,* (2009 Edition, by London: Bloombery). First published in 1980, by Writers & Readers Publishing Group, 65.

> take it in both its unity and its complexity, and endeavour to integrate all the special values in one fundamental value.[31]

The connection of what *house* and *home* means to me made it important for the research study to base the phenomenological and photographic experience Nettleham Hall.

8 On Location: Nettleham Hall

Through the course of observing and photographing various spaces in and around the hall, four areas in particular became central to the experience of absence and photographic inquiry: the dining room, front entrance, kitchen ceiling and the narrow path that leads to the hall. Through regular reflection, I discovered that I kept returning to these spaces. These areas resonated a particularly strong sense of awareness of absence and spatiality. In the subsequent chapter in this volume, Christobel Kelly expands on the importance of choosing a particular area of Australian landscape in which to take a 'long walk in order to create a visual and investigative response to this space.'[32] The reason for returning to specific spaces in and around the location of the hall was my interest in what these particular spaces were used for and the connection they offered. Interestingly, the areas of the hall that became central to the experience of absence and photographic inquiry were in essence the public spaces. These areas, in particular, were a regular cycle of activity for the inhabitants of Nettleham Hall. Similar in one respect to the resonance portrayed in the photographs of public spaces at *Chernobyl* by Robert Polidori and the public spaces of *Pripyat*, by Jane and Louise Wilson, they act as metaphors of past lives and traces of people. There is a real sense of the texture of absence in these spaces and existential presence. Experiencing and filming these spaces allowed me to consider the value of the art installation, which would connect the physical site of the hall to the experience of it. Scale became an important factor. To feel the connection of being there in the space, it became essential that I would need to produce large-scale outputs that were approximate in size to original the space. Throughout this process, the camera has acted as an extension of my body response to the experiences at the hall as I have embodied

31 Gaston Bachelard, *The Poetics of Space*, trans by Maria Jolas. (Boston: Beacon Press, 1994), 3.

32 Christobel Kelly, 'Colonial and Gothic Dissonance: Tracing Desire Lines Through Postcolonial Landscape,' in this volume.

them. Resonating with Muramoto's[33] analysis of Merleau-Ponty's body-object connection, the camera has acted as an observational and analogous response to the textured experience of being in and around the ruined space of the Hall. To further locate photography as an extension of body response and a communicator of experience in the ruin, the resulting art installation further asserts the lived experience as an interconnecting part of the whole.

9 The Art Installation and Its Significance

This section of the chapter is a reflective commentary of the art installation, which took place at Hackthorn Village Hall. It acts as a final response and embodiment of the observations and photographic responses at Nettleham Hall. It will first address the meaning and nature of the art installation before locating and contextualising it as a continuum of the experience and photography of absence for this research study.

What is art installation? Art installation, involves the installation or placement of objects in a space, where the entirety of the objects within that space comprise the artwork. Usually installed on the site where it will be exhibited, art installations can incorporate physical features of the site or simply exist within the space of the chosen site. The significance of the art installation is in the experience that it offers the viewer. Installation artist, Bonnie Mitchell suggests that 'installations take into account viewer's entire sensory experience. Installation art reflects and responds to the world we live in.'[34] Ronald Onorato further suggests in the book: *Blurring the Boundaries: Installation Art 1969–1996* that

> the aesthetic power of installation art does not reside in a singular, commodified object but in an ability to become, rather than merely represent, the continuum of real experience by responding to specific situations.[35]

Reacting directly to photography as a phenomenological inquiry into an experiential encounter, Jane and Louise Wilson's corresponding art installation,

---

33 Muramoto, in this volume.

34 Bonnie Mitchell, *The Immersive Artistic Experience and the Exploitation of Space*, (CAT, London Conference, February 3 2010), (viewed 15 January 2015), 101 http://www.bcs.org/upload/pdf/ewic_ca10_s3paper2.pdf.

35 Ronald J Onorato. *Blurring the Boundaries: Installation Art 1969–1996*, (San Diego: Marquand, 1997), 8.

A Free an Anonymous Monument, consists of both still and moving images that are multi screened and multi dimensional in size. The project itself is about the legacy of modernist architecture set within a post-industrial landscape of North East England and also conveys the psychology of architectural and technological waste. The installation is carefully choreographed to convey a complex and troubled history of urban regeneration.[36] The multiple, large-scale screens display looping film pieces and still photographs at differing angles of juxtaposition, height and size. A recurring feature of Jane and Louise Wilsons' work is that photography, as an observational tool, connects architecture to spatial experience and time. The installation of their work often includes multiple points of reference and experience – it becomes an essential interconnecting part of the practice.

In a similar fashion, photographer Fiona Crisps' large-scale installations explore the relationship between the phenomenology of photography and the space experienced. *Negative Capacities: The Stourhead Cycle, 2012,* was an English Heritage commissioned work, where Crisp was asked to explore the nature of the historic house and gardens of Stourhead in Wiltshire. The installation removes the gallery as a traditional plane of a walled exhibition of photographs and instead positions each of the eight, large-scale glazed and framed photographs on a single scaffolding pole within the space of the gallery. This deliberate action by Crisp created a new, temporary architecture of the gallery interior and allowed the viewer a more spatial and visceral experience of the work. The connection between the installation and practice was to offer the spectator a new way of looking and experiencing the house and gardens of Stourhead.

The significance of the art installation that grew out of the Nettleham Hall project is alluded to in the previous subsections in this chapter where photography is located as an analogous response to absence as a felt experience. The camera acted as an experiential extension of body experience. The art installation has the capacity to further embody this by drawing together of the interconnected parts that were made at the location of Nettleham Hall. *The Dining Room* film piece, *The Front Entrance* still photograph and the *Path to the Hall* culminated in a holistic environment in which the experience of absence further exists and unfolds. Analysis of the meaning and significance of art installation places it as an essential inter-connecting part of the whole to the experience of absence at the ruin of Nettleham hall. The following documents

36 Guiliana Bruno, *Modernist Ruins, Filmic Archaeologies,* 2003. Lost Sites of Power, Ruins: A Document of Contemporary Art, edited by Brian Dillon (Cambridge: MIT Press 2011), 76.

the meaning and significance of the art installation as a dissemination of a photographic and phenomenological inquiry for this research study.

Photography, as an act of making and creating, developed and responded directly to the experience of my time at Nettleham Hall, to the body awareness of the environment and the absence encountered there. An awareness of the art installation as a continuum became an essential part of the observation of the Hall and practice as it occurred. The choice of location for the installation was carefully considered in the planning of environment and large-scale works. It felt necessary to install the work in a space that considered natural light, existing architecture and height. This approach to the location allowed me to avoid the *white cube* associated with the traditional gallery space and open up the possibilities for spatial experience.[37]

10 Hackthorn Village Hall: The Location for the Installation

Having had previous knowledge of the space at Hackthorn Village Hall, I felt that the installation would work here. It is also located only ten minutes away from the ruin of Nettleham Hall and so holds certain significance for me. The village hall is a converted barn and offered the height and size that I would require. Inside the main hall are two windows located high above the floor and offered the opportunity to utilise or obscure. The space is used regularly as a village hall. There is an existing small stage at one end and signage throughout, which one would expect to find in a publicly used venue. The reuse of this space as a temporary place for the installation of Nettleham Hall became challenging at times yet afforded a strong sense of how any existing space can be transformed into an entirely different experience.

The Path to the Hall: It felt necessary to create a journey or path leading to the hall where the journey is as much part of the experience of absence as the hall itself. This element of the installation has been the most problematic component of the process. The original intention was to create a narrow, self-supporting *tunnel* made of white stretch fabric and using a series of projectors located on the exterior of the tunnel, project still images of the woodland onto the fabric tunnel. After testing this it was immediately apparent that the fabric of the tunnel was too heavy, not just in terms of physical weight but, crucially, it did not offer an analogous experiential feeling of walking through the

37 Ronald J Onorato. *Blurring the Boundaries: Installation Art 1969–1996,* (San Diego: Marquand, 1997), 16.

woodland path. Many further attempts to create a path to the hall were made using different materials, such as bamboo, willow, branches and twigs. Finally, I decided on cutting the existing white stretch fabric originally intended for the tunnel into thin strips of approximate lengths equal to the height of the tree canopy of the original path at Nettleham Hall. These strips were then tied at one end to a thin web of green braided line. The web represented the canopy of the woodland and the tied ends of the fabric strips acted as further representation of the canopy. The fabric strips were placed next to each other in parallels of three rows to create some density of woodland but also allowed gaps. It was essential to the experience that the viewer was allowed to peer through the gaps at any point and into different spaces of the installation. This also allowed the viewer to interact with the path and diverge off the path by walking through the small gaps between the fabric strips at any point. The path became an ephemeral trace of the original path and so provided the rationale for using the lightness and weightlessness of white fabric strips.

The path required some form of lighting to represent the dappled sunlight that appears intermittently in sections of the original path. One small lamp was used in a far corner of the room just outside of the immediate course of the path. The second light source, opposite the first light source was gained by utilising one of the windows from the exterior of the building. The window was partially covered with dark material, which allowed for movement of wind and intermittent sunlight to pass through the window and onto the path. At the end of the path the space opens up to the four suspended fabric panels that represent the front entrance to Nettleham Hall.

The Front Entrance: At the time of experiencing and photographing the front entrance it felt necessary to represent the still image acquired as approximate in size as the original front entrance of four meters in height. As well as creating a phenomenological experience of absence and space, the image itself was also intended to act as an entrance and therefore required consideration in its delivery. I found that digital printing onto fabric was a possibility and offered a way to present the entrance as an entity as well as an entry point. Through trial and error of printing onto different materials, the sheerest and lightest fabric available offered the result needed for installation. The outcome was the still image of the front entrance cut into four equal, four metre vertical panels. The reason for separating the image into separate panels was to allude to the *interior* and *exterior* experience of the hall as referred to at the beginning of this chapter. The panels were each suspended from transparent fishing line to the ceiling next to each other and a foot apart in distance. This made the image appear as a single image from one angle and fragmented from other angles.

The distancing offered the viewer different ways through the entrance much like the multiple ways of entering the original space. Viewing the panels from the front created a fragmented image due to the large-scale film screen behind the panels. Parts of the image on the panels appeared opaque, whereas other parts appeared translucent. This deliberate action alluded to the fragmentary and fragility of the original structure as well as a sense of absence. Using sheer fabric, the panels could be viewed from the reverse. A more opaque view of the entrance was also created by the low density of light caused by the partially covered window and single lamp that was directed onto the path. By suspending the panels in the centre of the space, the viewer had ample room to walk through and around the panels obscuring the line between installation and real experience.

The Dining Room: The dining room was filmed as a static piece of footage with the intention of looping the film. This allowed for a continuous narrative of absence as it was being experienced. Similar to the approximate scale of the *Front Entrance*, I also felt it necessary to apply the same sense of scale to the film piece. Therefore, a large wooden frame was constructed and painted white to house the 5 metre by 3 metre fabric screen made from the same material as the path. The film was projected from the rear removing traces of technology required for this part of the installation.

The looping of the film corresponded with the audio of nature specific to the environment of Nettleham Hall. Our awareness of the senses of sound in particular enriches our experience of the world around us. Through our hearing, sounds in architecture such as the richness of amplified sound in a theatre 'can increase the intensity of its perception.'[38] It felt necessary to introduce the natural sound of the environment of Nettleham Hall to the installation experience rather than introduce audio that is not associated with the hall. The sound, made by specialist equipment, was conducted at the time of my encounter with the hall and is made up predominantly of birdsong local to that area. The recording was made during early mornings of the summer over a period of two years. It resulted in a rich array of birdsong and additional sound of wind, trees and small animal activity. The inclusion of this environmental sound was crucial in drawing together the inter-connected parts of the experience into an entirety.

38 Panagiotis Hadjiphilippou, 'The Contribution of the Five Senses Towards the Perception of Space.' Academia.edu, (January 2013). Chapter 4: Sensual Projects. (Viewed 20 January 2015) https://www.academia.edu/2460561/The_contribution_of_the_five_human_senses_towards_the_perception_of_space_by_Panagiotis_Hadjiphilippou.

11 The Installation as an Entirety

On entering the open, double doors to the main room of the Hackthorn village hall one was confronted by a series of white fabric strips hanging from high above the head to the ground and the sound of birdsong. There was a narrow path to the immediate right created by countless vertical suspended strips. The path wound its way along. There were small gaps here and there through the strips suggesting alternative ways and views into the space beyond. The birdsong was louder halfway through the path and the intermittent light from above and below created a dappled effect all around. The vertical strips that comprise the path were soft to the touch. At the end there was a wide opening that lead to four vertical panels of sheer fabric suspended from the ceiling. The image on the panels consisted of the front entrance to Nettleham Hall. There was space to walk through and around the panels experiencing a new sense of scale, absence and a spatial awareness of the environment as whole differing from that of the path.

Beyond the vertical panels was a large-scale screen using the full width of the room. The film piece was a static view of the dining room. It was presented as an entire section of the dining room, where the floor of the room is in line with the floor of the installation space. The film piece of the movement of ivy and dappled sunlight was on a continuous loop juxtaposing the stillness of the fragmented architecture of the room.

The combination of film footage, still image, audio and path created an experience of 'being there' in the ruin of Nettleham Hall and, yet, the absence is felt through the body and awareness of the senses. The experience of the installation in its entirety recreated the space of the Hall through its absence.

12 Phenomenological Possibilities of Photography

The original motivation driving this research was to explore the phenomenological possibilities of photography. As a practitioner of the medium, I am interested in how photography interacts and reacts to experience. A long-standing fascination with the ruin prompted the two elements of practice and place to form a framework this research study. The aim of the research was to investigate how the ruin affects us in an experiential way, how our bodies respond to the ruin and absence and how that occurrence can then be visualised through photography into a tactile and holistic experience. A Phenomenological approach was adopted to examine the core elements of inquiry: absence as a felt experience, photography as an extension and visualisation of body experience and, finally, art installation as a continuum. These parts intervolved into an artefact of inter-connecting parts. In drawing together the elements through phenomenological elements, it was essential to first analyse the meaning and significance of each one in turn for the purposes of understanding their values to this study before conducting experiential investigations. The research question was divided into two reflexive and inter-connected parts that aimed to more fully identify the elements that the research posed. The following summary recognises that the research study is broadly interpretative but outlines the importance of phenomenology as an inquiry – to uncover the meaning of something it is necessary to experience it, as suggested by Merleau-Ponty.[39]

In the first part of this chapter, absence was observed as a *felt* experience, it was then interrogated as to how it occurs within the spatial setting of the ruin. It was suggested that absence is perceived through the body and awareness of the senses. The perception of space involves the body as an entirety. The connection between sensory information and emotional response enable us to react and interact with the ruin. As suggested by Trigg, an experience of absence particular to the ruin is offered in a very real, tactile and textural nature.[40] The experience of absence was then located to a specific spatial setting, the ruin of Nettleham Hall. In the form of an ethnographic journey the Hall was phenomenologically situated as the location for the experience of absence and photographic response.

A phenomenological inquiry examined the significance of photography for this research study and explored the importance of the medium as practiced by others. Photography was shown to have the capacity to contribute to a heightened sense of experiential quality of the ruin through capturing and recording duration of time. In favour of this, Susan Sontag refers to photography's phenomenological ability to capture experience and that the camera acts as an extension of the body.[41] From this point of orientation, selected works by photographers' Jane and Louise Wilson, Hiroshi Sugimoto and Robert Polidori, were examined and positioned as photographs that acted as signifiers of phenomenological experience. The photograph was recognised as an embodiment of lived and spatial experience.

39 Maurice Merleau-Ponty, *The Phenomenology of Perception,* trans by Colin Smith (London & New York: Routledge, 2006) 282.

40 Dylan Trigg, 'Architecture and Nostalgia in the Age of Ruin,' (Viewed 7 December 2014), 2. https://www.academia.edu/208447/Architecture_and_Nostalgia_in_the_Age_of_Rui, (accessed 7 December 2014), 2.

41 Susan Sontag, *On Photography* (New York: Penguin Modern Classics, 2002), 22.

In its widest sense, photography includes the moving image and its similar phenomenological position to the still image. The analysis of the film piece *Kodak 2006* by Tacita Dean provided further reference to the ontological nature of photography as a phenomenological response to its making and subsequent viewing. Dean's film piece offers an effective use and treatment of photographic filmmaking that is significant to translating spatial experience. The film also touches on the notion of ruin and how one can communicate a phenomenological experience and successfully interpret that experience into a film piece.

The core elements of this research served to position the practice as a continuum of phenomenological inquiry. Providing ethnographic accounts of Nettleham Hall as a ruin, the practice as it occurred there and the installation as a response add to the understanding and significance of the phenomenological value that this research is based on. Additionally, this approach aims to give the reader an essence of the experiential encounters as they happened. The observational encounters at Nettleham Hall directly influenced the photography that was made there. This Empirical approach allowed myself, the subject, as the primacy of perception and the camera an extension of that perception. A combination of still and moving image was made at the Hall responding to my reactions and interactions with absence. Audio sound was recorded in addition as an extension of the experience there. It is important to point out that the photography made on site was also influenced by the eventual art installation. This provided a more holistic and Hermeneutical approach to the process. It recognised that the whole is reflected in the parts just as together the parts reveal the whole.

Finding a suitable venue for the art installation was paramount. It felt necessary to maintain the characteristics of exposed architecture and space as well as consideration of the scale of the final works. Hackthorn Village Hall provided the space and tactile nature the installation required. The venue was offered for a period of four days. This short space of time, in which to erect, show and disassemble the installation, initially presented a multitude of concerns. The short length of the installation on one day and an evening, however, was in favour of the nature of the ruin and refers to the title of the installation and this subsequent chapter as: *Perpetual Transitions*. The constant change that the ruin faces and truth of entropy is highlighted with a *Pop Up*[42] style installation.

42 The feature of the *pop-up* exhibition is that they come and go in a short space of time, and typically appear in places not usually associated with art exhibitions. They can possess an element of surprise for the viewer, altering the experience of seeing it. Pop-up galleries often emerge in empty spaces, or spaces used for other purposes. They expand the available space for artists to show their work, and give artists a curatorial control. (Sally Pollak, *Pop-Up*

During the opening of the installation to the public a number of surprise elements occurred. I was interested to see how the viewer reacted and interacted with the space of the installation. A number of viewers did not choose to walk through the path instead they found other ways through. This observation defines and highlights how differently we each consider space. The installation was presented in such a way that the viewer had multiple ways of interacting and moving around the space. I observed viewers sitting on the floor in front of *The Dining Room* projection. In addition, many viewers touched the fabric strips of *The Path* as they walk through as well as the fabric panels of *The Front Entrance*. These observations are critical in outlining the necessity for the photography conducted at Nettleham Hall to be presented as an installation, a continuum of the felt and tactile experience.

13 Conclusion

In closing, this research study presents the value of the phenomenological inquiry, of a philosophically based study and how they can extend to and encompass practice and dissemination. This research also makes recommendations for further inquiries into the phenomenological field of photography as an experiential inquiry. In the process of this study, a significant amount of photographic material has been made at various ruined sites including urban and industrial. Further experiential investigations into these spaces will provide a deeper examination of absence, photography as a phenomenological inquiry and the art installation as a continuum of that experience.

Bibliography

Bachelard, Gaston. *The Poetics of Space.* Translated by Maria Jolas. Boston: Beacon Press. 1994.

Barthes, Roland. *Camera Lucida.* London: Vintage. 2000.

Berger, John. *About Looking.* London: Bloomsbery. 1980.

Bruno, Guiliana. *Modernist Ruins: Filmic Archaeologies.* Lost Sites of Power. Ruins: A Document of Contemporary Art. Edited by Brian Dillon. Cambridge: MIT Press, 2011.

Galleries: a Growing Trend in Burlington's Art Scene, Burlington Free Press.com, March 2013) (Viewed February 27 2015) http://archive.burlingtonfreepress.com/article/20130331/ARTS/303310001/Pop-up-galleries-A-growing-trend-in-Burlington-s-arts-scene.

Concise Oxford English Dictionary. 'Ruin.' 12th Edition. Oxford: University Press, 2011a.

Concise Oxford English Dictionary. 'Sensory.' 12th Edition. Oxford: University Press, 2011b.

DCA Dundee. 'Interview with Jane and Louise Wilson,' YouTube. Online video clip, (viewed February 27 2015) http://www.youtube.com/watch?v=Wm4vl6DhUXY.

Duråo, Maria Joao. *Embodied Space: A Sensorial Approach to Spatial Experience,* Propulsion & Energy Sciences International Forum: SPESIF-2009. AIP Conference Proceedings, Volume 1103. AIP Conference Proceedings, Volume 1103, Issue 1. (viewed 15 January 2015) http://www.astrosociology.org/Library/PDF/Durao_SPESIF2009.pdf.

Dillon, Brian. *Ruin Lust.* Catalogue for the exhibition *Ruin Lust,* Tate Britain, 4 March to 18 May, London: Tate Publishing, 2014.

Dillon, Brian, introduction to, *The Ruin: Documents of Contemporary Art,* edited by Brian Dillon. Cambridge: MIT Press, 2011.

Hadjiphilippou, Panagiotis. 'The Contribution of the Five Senses Towards the Perception of Space.' Academia.edu. January 2013. Chapter 4: Sensual Projects. (Viewed 20 January 2015) https://www.academia.edu/2460561/The_contribution_of_the_five_human_senses_towards_the_perception_of_space_by_Panagiotis_Hadjiphilippou.

Kraucer, Siegfried. *The Theory of Film: The Redemption of Physical Reality.* Oxford: University Press, 1960.

Kuhn, Annette and Kirsten Emiko McAllister. *Locating Memory: Photographic Acts, Remapping Cultural History.* Volume 4. New York: Berghahn Books, 2006.

Merleau-Ponty, Maurice. *The Perception of Phenomenology.* Translated by Colin Smith. London: Routledge, 2006.

Mitchell, Bonnie. *The Immersive Artistic Experience and the Exploitation of Space,* (CAT, London Conference, February 3 2010), (viewed on 15 January 2015) http://www.bcs.org/upload/pdf/ewic_ca10_s3paper2.pdf.

Onorato, Ronald, J. *Blurring the Boundaries: Installation Art 1969–1996.* San Diego: Marquand, 1997.

Roth, Michael, Claire Lyons and Charles Mereweather. *Irresistible Decay: Ruins Reclaimed.* Los Angeles: Getty Research Institute Publication, 1997.

Sontag, Susan. *On Photography.* New York: Penguin Modern Classics, 2002.

Trigg, Dylan. *The Aesthetics of Decay: Nothing, Nostalgia and the Absence of Ruin.* New York: Peter Lang, 2006.

Trigg, Dylan. *The Memory of Place: A Phenomenon of the Uncanny.* Ohio: University Press, 2012.

Trigg, Dylan. 'Architecture and Nostalgia in the Age of Ruin,' (Viewed 7 December 2014). https://www.academia.edu/208447/Architecture_and_Nostalgia_in_the_Age_of_Ruin, 2.

Wollon, Peter. *Fire and Ice.* In Photographies, Number 4, Paris, 1984.

CHAPTER 10

Colonial and Gothic Dissonance: Tracing Desire Lines through Postcolonial Landscape

Christobel Kelly

Abstract

The space of Australian postcolonial landscape has been fashioned by pragmatism and inured with the cultural experience of displacement, incarceration and the persistent need to lure free settlers from Europe. Out of these early settlement imperatives a style of representation arose that arranged the landscape as an object of culturally dissociated contemplation, figured by forgers in order to transform the view into something idyllic, tame and acceptable to the English eye.[1] Another divergent style of representation borrowed from literature, however, depicts Australian gothic landscape as 'a world of reversals, the dark subconscious of Britain.'[2] In this landscape we encounter a haunted backdrop for settlement wherein 'immeasurable isolation'[3] combined with desperate uncertainty and unspecific danger in an unfamiliar place.[4] More recently Australian visual artists have used immersive experience as a way of establishing a close proximity to landscape. However, in doing so, there is always an awareness of 'an echo from our colonial past,'[5] which demands attention. In this sense, postcolonial landscape may also be seen as a ruin where elements of Australian gothic sensibility characterised by desolation, secrecy and isolation, create a disconsolate space. Postcoloniality therefore generates a dilemma for contemporary Australian artists who may grapple with landscape that is imbued with absence.

---

1 Robert Hughes, *The Fatal Shore,* (London: Pan Books, 1988), 599.

2 Gerry Turcotte, 'Australian Gothic,' in, *The Handbook to Gothic Literature,* ed Mulvey Roberts, Basingstoke: Macmillan, (1998), 10. Even before Australia was explored and mapped, it existed as an imagined place peopled by monsters, and its settlement as the dungeon of the world, confirmed this Gothic sensibility in an era of sensationalist literature.

3 Gregg, Simon, *New Romantics Darkness and Light in Australian Art,* (Melbourne: Australian Scholarly Publishing, 2011), 71.

4 Turcotte, 'Australian Gothic,' 1.

5 Mimi Kelly, *Stranger in a Southern Land,* (Sydney: A-M Gallery, 2014), 9.

 | DOI 10.1163/9789004365520_012

1 Inserting the Body into Time and Space

In 2013, with the idea of close proximity to Australian landscape in mind, I undertook a long walk in order to create a visual and investigative response to this space. The area I chose as the primary resource for my art practice was not some remote expanse. Instead I selected the place where I had grown up and lived for the whole of my life, [1] a parochial stretch of land, which runs from my hometown of Mylor in the eastern Mt lofty ranges, down through the Fleurieu Peninsula to the costal spot of Second Valley. The end point of my journey was a small blip of a town which sits on the ankle of the Peninsula whose foot kicks out into Gulf St Vincent within striking distance of the roughly football shape of Kangaroo Island.

The purpose of the creative journey was an interrogation of the walked landscape. In his essay 'The Known World' (2010), cultural theorist Ross Gibson examines how tacit knowledge accessed in the creative arts is used as a foundation for scholarly research, that is, how the cognitive mode operating when one is making art can be interrogated to give a clear account of the process.[6] Gibson explains that the artist is an experimenter who, after identifying a research question or aim, collects information and then formulates methodical experimentation as a way of stepping away from what they already know and into the mystery of their topic. Gibson continues, 'the researcher flows along with the dynamics and complexities of action and reaction within the mystery until some shift in understanding has been sensed or observed.'[7]

I intended, therefore, to immerse myself directly in landscape by following the trace of 'desire lines,' a term borrowed from the field of architecture and urban design, which refers to the path of least resistance across walked terrain. As the writer Robert Macfarlane explains, the paths taken by the walker often follow visible traces of the passage of humans and animals who leave their marks on dirt, sand, mud, moss, grass and frost.[8]

In the following chapter in this volume, Anthi Kosma also suggests that the body itself is a generator of space.[9] Accordingly, the trace of desire lines could be regarded as a cumulative act of drawing or writing on the surface of the land, an exteriorization or inscription left as a residue to be read almost as

6 Ross Gibson, 'The Known World,' *Text: Symposium: Creative and Practice-led Research*, (2010): 1.

7 Ibid., 5.

8 Robert Macfarlane, *The Old Ways*, (New York: Viking, 2012), 13.

9 Anthi Kosma, 'Spacing by trace: an event of an out of place experience,' 2014, in this volume.

'an archaeological excavation.'[10] Thus the desire lines I followed existed as a result of the movement of bodies through space. The methodological approach to my project was based on action research that involved a series of activities and observations, which were made and then reflected upon.[11] The idea of walking the landscape along 'desire lines' in order to understand the discordant notes of colonial and gothic landscape for instance, was the mystery uncovered incrementally by perceptual engagement.

And so for this long walk I headed south by stepping into a mutable space that was shaped by topography, capricious atmospheric conditions and pathways formed by movement of the animate world, a space where tricks of light and changing weather transformed the ordinary, familiar landscape into one that was almost unrecognisable. Underlying this project, however, were troubling concerns about Australian contemporary relationship with landscape, worries which had asserted themselves as amorphous dissonant queries. One part of the query was concerned with how landscape had been understood in a post colonial sense and another part with how contemporary artists reconcile themselves with landscape as a lived space that is contested, atavistic and contemporary all at the same time.

The length of the journey was about 95 kilometres and I had previously walked much of it in sections along the Heysen Trail.[12] There were, however, unexplored gaps along the way and so intense was my longing to do undertake this journey that these spaces existed in my mind's eye as an *ekphrasis,* meaning an excessively imagined subject.[13] I did not have a map and the unfamiliar sections, I believed, were navigable in the same way that I had traversed the more familiar sections. As the contemporary art theorist Lucy Lippard explains, maps often leave out features that are important to the walker. Trees fall and grow, paths wash away and fire razes the biosphere, all of which demonstrates that landscape itself is in a constant state of topographical transformation.

10 Kosma, in this volume.

11 'Action Research Guide for Alberta Teachers,' Alberta Teachers Association: 2000, viewed 14 February 2015, http://www.teachers.ab.ca, 12.

12 Warren Bonython, 'History of the Heysen Trail,' viewed 14 February, 2015. http://www.heysentrail.asn.au. This 1,200 km South Australian walking trail stretches from the Flinders Ranges down through the Adelaide Hills and was named after the painter Hans Heysen who painted in both these areas and was an energetic walker in pursuit of subject matter for his work.

13 Camille Guthrie, 'Writing Poetry About Art', *Poetry foundation*, viewed 20 May 2014, http://www.poetryfoundation.org/. This refers to a vivid scene or work of art where imaginative narrative magnifies the meaning of the described object.

On a more personal level, 'the depth of individual emotion engendered by place is also unpredictable,'[14] and I had a complex associative relationship with this area. The first ten years of my life I had lived in the middle of this elongated tract in the sleepy town of Aldinga where I had walked to and from the beach and school under rows of gnarled and resinous pine trees. Those walks are now almost unrecognisable due to creeping urbanisation although sometimes there are vestigial memories that rise up when I travel back. These are places I now inhabit in my dreams. Altered, reconfigured and sometimes unreachable, they have become part of the geography of my psyche, a landscape that for me is loaded with and codified by memory. My relationship with this place forms what Lucy Lippard describes as 'reciprocal identity,'[15] wherein one is shaped by the landscape and in turn one shapes the landscape.

And so on the first day of the walk I set off in the pouring rain and despite having an intimate knowledge of the area, I was soon lost. Backtracking for an hour or so till I found my way forward, I reflected on how it was that others had traversed this space. In particular how colonial exploration had impinged on the landscape.

2 Colonial Exploration

Early Australian colonial expeditions took their intellectual and cultural cues from Europe where expanding interest in knowledge of the natural world was used by travellers as a way of appraising the natural features of landscape.[16] The collection of scientific data on voyages of discovery during the eighteenth and nineteenth centuries had a provocative effect on accepted theories about the natural world. Artistic, literary and philosophical practices moved towards an understanding of 'nature' and 'natural' as entities that existed beyond human control or explanation.[17] Consequently, when the colonial explorer Surveyor General Sir Thomas Mitchell (1792–1855) used notions of the grand and terrible in the assignment of place names, he sought to find the romantic

14 Lucy Lippard, *The Lure of the Local Senses of Place in a Multicenterd Society,* (New York: The New Press, 1997), 79.

15 Lippard, *The Lure of the Local,* 6.

16 Julia Horne, *The Pursuit of Wonder, How Australia's Landscape was Explored, Nature Discovered and Tourism Unleashed,* (Melbourne: Melbourne University Publishing, 2005), 34. Before 1830 the term 'expedition' was used synonymously with 'tour' and 'excursion,' and after that, 'expedition' was favored as a descriptor of journeys of discovery. By 1850, however, 'exploring' and 'exploration' dominated accounts of those trips.

17 Ibid., 36.

sublime in the Australian landscape and hence imbue it with meaning and allowing the countryside to be viewed as a new version of home. In traversing the physical space of the land the explorer recruited place names such as Mount Salvator, the Claude River, Martin's Range (for the painter John Martin) in order to inscribe landscape with the imagined possibility of familiarity and settlement.[18] Mitchell bucolically positioned himself as an enabling presence in the landscape by quoting classical text in his journals and linking himself to the Roman Poet Ovid,[19] 'and the ground which had hitherto been a common possession like the sunlight and air, the careful surveyor now marked out with the long-drawn boundary line.'[20] The subtext to this imposing nomenclature was a sense that Mitchell believed any consideration of the area as Aboriginal was easily dismissed and that he had encountered a dehistoricised space.[21]

It follows that Mitchell's deliberate coopting of the names of visual artists for place names was a way of framing what features were seen in landscape. As Shona Hill quoting Toffoletti explains, 'it is important to remember that ... technologies- such as painting and photography – have influenced the way we have come to understand ourselves.'[22] We understand the 'sense' of an image through a bodily response even before we 'cognitively process their meaning and morality.'[23] As a result, this legacy of ordering and objectifying the physical features of landscape within a frame shapes our cultural connection to the land.[24]

Subsequently, Australian artist Mandy Martin's large-scale works *Peripecia, the Salvator Rosa Series* (2002) explore that same landscape investigated by Mitchell, which has a multiple past. Drawing on extracts from the explorer's journals and fleeting meteorological apparitions, she overlays the paintings with texts that evoke the classics such as Ovid's *Metamorphosis* (1ACE). This can be seen as an inscription of the 'layering of landscape with its previous ... *cultural* history.'[25] In this way Martin demonstrates how place names are

18 David Malouf, 'Peripecia,' in *Mandy Martin: Peripecia, the Salvator Rosa Series,* (The Australian National University: Canberra, 2002), 12.

19 Ibid., 13. Taken from, *Metamorphosis* (AD 8) Ovid's epic poem consists of a number of discreet but interlinking story lines concerning changeability.

20 Ibid., 13.

21 Eric Rolls, *A Million Wild Acres,* (Melbourne: Penguin, 1984), 98.

22 Shona Hill, 'Imprinted on the Body: Changing Sensibilities in Time and Space,' paper presented at the '4th Global Conference on Time, Space and the Body.'

23 Ibid., 1.

24 David Malouf, 'Peripecia,' 11.

25 Ibid., 11.

also signifiers of exploration and journey in that they trace the way the spatial landscape impresses itself on the consciousness of the explorer or surveyor.[26]

By day two of my own explorative walk the sky was beginning to clear and the thinning clouds cast shadows on the rolling hills as I followed sheep paths and kangaroo tracks. Parallel to my path was the subtle gouge of an old coach road smoothed over with grass. Perhaps I should have carried a map and entered into a Cartesian model of gridded space. However, the Australian cultural theorist Russel Smith argues that abstracting space into a mathematically representable format is antithetical to social relationships and 'the bodily and experiential dimensions of travel.'[27] From the discussion above it follows that in some way, the intensity of uncertain navigation ensured that I was compellingly engaged with the landscape.

The walking project, therefore, drew on Eric Mathew's *The Philosophy of Merleau-Ponty* (2002), in which he explains Merleau-Ponty's idea that if we were to look at the world in a purely scientific manner, which exists independently of ourselves, we would in some way set aside our place in the world. The world is, therefore, not something we merely look at from an outside position but something we occupy with our incorporated minds. Furthermore, the foundation of our experience of being in the world comes about through perception and not through a distanced or overly objective reading of existence.[28] Thus phenomenological methodology allowed for a complex alignment of perceptual elements wherein meaning was attached to particular events such as navigation by desire lines, the reciprocal identity experienced in this particular landscape or the flow of time when walking.

By day three it became apparent that I had begun to walk backwards in time. If one thinks of the shape of this journey as a ribbon draped across the land from Mylor to Second Valley, more or less tracing the wide groove of an extinct glacier, then near the middle of the ribbon is the town where I came into the world. Subsequently, as I moved forwards on day three of my journey I also moved into the epicentre of remembered landscape.

Growing up in the drowsy town of Aldinga in wheat-field country south of Adelaide I had understood that the name meant place of wild ducks. Aboriginal people had bestowed the name in some distant past. There were, however, neither ducks nor Aboriginal people there when I was a child. The unsettling

26 Paul Carter, *The Road to Botany Bay*, (London: Faber and Faber, 1987), 30.

27 Russell Smith, 'The Spatial Poetics of Computer Games and the Mapping of Australian Space,' *Halfway House the Poetics of Australian Spaces*, Jennifer Rutherford and Barbara Holloway (eds) (Western Australia: University of Western Australia 2010), 353.

28 Eric Mathews, *Philosophy of Merleau-Ponty*, (Durham: Acumen Publishing, 2002), 47.

idea that they may have been there once or that the denuded land had ever been home to flocks of ducks gave the landscape a vestige of emptiness. This was a space that had been inhabited by others and over time had evolved to contain a small town dwarfed by the spreading paddocks, a long arc of hills and the wide sinuous coastline that offered up glimpses of the mast of the shipwrecked 'Star of Greece' at low tide. In the absence of any tangible 'other'[29] I had peopled the landscape with characters, ghosts and stories I gleaned as an inveterate eavesdropper. Sponge-like I had absorbed literary narratives and transposed this imagined world into the real landscape.[30]

Facts, contemporary artist Jill Orr clarifies, can only explain a certain amount before they become dry and distancing and humanity flourishes whilst contemplating the imaginary 'where stories, narratives and shared cultural expression can be the spark that energizes multiple forms of creative manifestation.'[31] What had percolated and risen into consciousness when walking through this landscape was a visceral memory of literature. As a child what could be simpler than replacement of the imagined world with the real world? Surrounded by absence – the missing from the wreck of the 'Star of Greece,' errant ducks, the long gone Aboriginal people, the tiny Cornish mud ruins at the Delabole slate mines – each of these haunting elements was slotted into a strange hybrid of culture and landscape where patterns of absence resonated with a sense of gothic landscape as a disconsolate space.

On a more personal level, by the time I finished this section of the walk near the Victory Hotel I was tired. I had been walking for three days, some of it in hostile weather. One foot was blistered where I had stepped in a creek and my wet sock chafed. On the other side my knee was swollen and painful. It was difficult to work out which side to limp on. I was also beset with nostalgia and an intense and unreasonable longing to move back to the area.

---

29 Kathleen Steel, 'Fear and loathing in the Australian bush: gothic landscapes in Bush Studies and Picnic at hanging Rock,' *Colloquy: Text Theory Critique,* (Melbourne: Monash University, 2010), 41. Australian colonial landscape stood outside cultural recognition and became imbued with otherness or that which lay outside the home. It was perceived as unstable and lacking in a history that would make it recognisable, thus becoming charged with both guilt and uncertainty.

30 Bruno Latour, *An Inquiry into Modes of Existence, and Anthropology of the Moderns,* trans. Catherine Porter, (Cambridge, Massachusetts: Harvard University Press, 2013), 251. Fictional narrative, Latour explains, can be thought of as a way of being carried away into another world where the beings of fiction act as emissaries, and bring back parts and multiples that can be unified and simplified to create a fuller picture of the imagined world.

31 Orr, Jill, 'Heat: Art, People and Climate Change,' *Transdiscourse 1: Mediated Environments,* ed. Andrea Gleiner, Angelika Hilbeck, Jill Scott. (Springer: New York 2011), 115.

3 Postcolonial Landscape as a Ruin

As visual artist Clementine Monro describes, there is a presence within absence, which can be felt as a spatial experience when one encounters an architectural ruin.[32] And so these spatial qualities of the ruin articulated by Monro could also be seen as encompassing postcolonial landscape. Over time, the landscape I had encountered as a child simultaneously contained both presence and absence, which had been sensed and intuited before it could be articulated. By day four of the walk when I found myself on the steep shouldered escarpment that led into the backlands behind Sellicks Hill, I had also begun to notice the intricately constructed dry-stone culverts designed to channel the water under the road. They were remnants from colonial times when this was the main road south. As Gilloch citing Walter Benjamin explains, objects on the point of ruination have an afterlife that reveals the complex relationship between old and new.[33] For Benjamin, the truth of an object was disclosed at the point where the context for its original existence had faded[34] and although this route had been abandoned long ago, the finely wrought box culverts suggested that they were almost too good for their current purpose.

Nevertheless they established a link to colonial times when in the push for settlement during the 1860s this was called 'The Victory Road,' so named after a long struggle with the authorities in Adelaide to create a road connecting the outlying towns of Myponga, Aldinga, Willunga and McLaren Vale. In those times, terrestrial roads were seen as a vital alternative to the use of more precarious coastal shipping. Shipwreck was an ever-present danger in costal traffic when moving stock and people into an area that was falsely considered vacant and ripe for the kind of British landed settlement that went hand in glove with the idea of subduing the earth.[35] However, by the time I came to walk there the road had an air of abandonment about it. It had not been surfaced and the stonework, drainage and culverts, I imagined, looked much as they would have in colonial times.

---

32 Clementine Monro, 'Perpetual Transition: A Communication Between Photography and Absence Within the Ruin,' in this volume.

33 Graeme Gilloch, *Myth & Metropolis Walter Benjamin and the City*, (Oxford: Polity Press, 1996), 14.

34 Ibid., 14.

35 Rob Linn, 'The Discovery and Settlement of the Fleurieu Peninsula and the Angas/Bremer Region 1802–1861,' *The Journal of the Historical Society of South Australia*, no.14, (1986), 56. http://www.sahistorians.org.au, viewed 12 August 2014.

4 Colonial Memory

Taking accounts from the time of European settlement, the critic Robert Hughes suggests that in the beginnings of colonisation Australians believed that they had encountered an 'historical blankness ... where nothing recognizable had happened for millennia.'[36] Consequently, as Australians understood that this was a de-historicised space they sought to retain British history as their own. Furthermore, as Hughes explains, almost all early landscape artists were forgers, counterfeiters or transported clerks.[37] Thus when the counterfeiter Joseph Lycett was encouraged by Governor Macquarie to produce a series of watercolours the landscape in *Views from Australia* (1824) became a kind of marketing ploy to lure free settlers. Hughes quips that 'There was something elegantly appropriate about setting a forger to such a task.'[38]

This pictorial reconfiguring, Anglophile yearning and the grimness of the convict experience have, according to Hughes, led to a collective desire for amnesia to do with the past.[39] This finding, however, becomes more nuanced when one scrutinises the genre of gothic landscape, which communicates itself through a complex set of modes 'delineated by borrowings and conflations.'[40] As an Australian literary form it often emphasised 'the horror, uncertainty and desperation of the human experience'[41] and represented the colonial landscape as the site of individual entrapment. Landscape, it seemed, was recognisable but unfamiliar enough to create a feeling of dislocation about the reversal of seasons and the strangeness of the fauna.[42]

5 Gothic Landscape

Hughes' assertion that colonial Australians imagined themselves alone in a de-historicised landscape is contested by the cultural theorist Kathleen Steel, who argues that in the struggle for settlement, anxiety concerning personal, local and societal history was a form of disquiet that became merged with

36 Hughes, *Fatal Shore*, 599.
37 Ibid., 339.
38 Ibid., 339.
39 Ibid., 597.
40 Turcotte, 'Australian Gothic,' 2.
41 Ibid., 1.
42 Ibid., 1.

the landscape, becoming the 'other'[43] or that which lay outside the familiar comfort of home. A literary reading of the merging of landscape with otherness can be found in Barbara Baynton's grim collection of short stories, *Bush Studies* (1902) as well as in Joan Lindsay's Gothic novel, *Picnic at Hanging Rock* (1967). Steel concludes that both books are informed by the denial of a pre-European history of Australia and, more importantly, that this denial is a fallacy. Where Baynton reacts to this notion by creating an atmosphere of desolation, Lindsay 'offers a nuanced awareness of the natural and Indigenous history of Australia.'[44] In their portrayal of landscape as a psychological space, Steel proposes that Baynton's stories highlight the landscape as a contested space where non-Indigenous civilization wrested the landscape from its previous inhabitants thereby creating a secret atmosphere of guilt and anxiety. On the other hand, the question of the girls' disappearance in *Picnic at Hanging Rock* can be read as a sacrificial offering that functions as an assuagement to a landscape, which is imbued with otherness. Thus there is a form of historical dissonance underpinning the notion of the Australian gothic landscape, which has implications for contemporary artists who work within the field of landscape as a psychological space.

From the above discussion it also follows that gothic landscape can be associated with the dread and anxiety in particular of becoming lost. The contemporary photographer Kurt Sorensen draws on elements contained in an eerie fragment of history in his series *I get some terrible frights here* (2014). The title taken from a 1918 newspaper article concerns the mysterious disappearance of David and Emily Jane Joel near Wentworth Falls in the Blue Mountains. After being seen intermittently in the days before, Emily is found exhausted and confused and later dies. Emily Jane's husband is never found. Sorensen's work plays into the fear of landscape as veiled in danger and otherness as well as highlighting the awesome beauty of the natural area.[45] This tenuous and uncertain link between fearfulness and admiration of natural surroundings creates a gothic tension in the depiction of Australian landscape, which resonated with my early memories of the area around Aldinga and Willunga that I had just walked through.

43 Steel, 'Fear and loathing,' 41. In the early struggle for domination, the landscape took on an entity that stood outside cultural recognition and became imbued with otherness or that which lay outside the home. Beyond the home, landscape was perceived as unstable and lacking in a history that would make it recognisable, thus becoming charged with both guilt and uncertainty.

44 Ibid., 49.

45 Kelly, *Stranger in a Southern Land,* 9.

For the contemporary photographer Jane Burton the powerful influence of Tasmania's formative geography is reflected in her work where folkloric history cloaks the landscape in an ominous air of mystery. Nevertheless, her work transcends geographical identification and 'is staged somewhere between dream and nightmare and employs the real in the service of the unreal.'[46] I reflected on the disturbing narrative in Burton's images and their connection to Sorensen's trope of fear of becoming lost. I pondered on loss and absence, remembering the shipwreck of the 'Star of Greece' (1888)[47] whose story had pervaded my early years and acted as a lodestar for gothic suggestion. The geographic area of the walk had drawn attention not only to the strange desolation of gothic landscape but also to the idea that this was a space where anxiety connected with absence in landscape had been in some way aligned with an unsettling denial of pre-European history.

6 Kaurna Country

Undoubtedly it is axiomatic that my understanding of Aboriginal connectivity to this landscape had evolved since my childhood and that the whole spreading country that I walked across was cared for and indeed has an Aboriginal history and presence. 'In Aboriginal cosmology' Cassie Plate explains, 'time past is part of the eternal now, in which the past goes on living in the present. It continues,' she says, 'with its layers of evolving stories and meaning into the future. Dreaming marks are permanent, they don't wash away.'[48] And so, my journey and the town where I lived as a child lay within the range of one of the best-known Dreaming narratives of the Kaurna people. This account has become closely associated with 'the lore and law of Kaurna country, and the process of contemporary political and cultural renewal.'[49] Part of this complex story tells the funerary journey of the ancestral being, Tjirbruke, who followed the coastline from the Adelaide plains to Rapid Bay carrying the body of his

46 Gregg, *New Romantics,* 65.

47 *Star of Greece,* http://www.abc.net.au/backyard/shipwrecks/sa/greece.htm, viewed December 2014. Laden with wheat, the ship was wrecked in a violent storm off the coast of Port Willunga. It was unclear how many people drowned due to discrepancies in the number of people on board but was thought to be no less than 18.

48 Cassie Plate, 'German Romanticism, British Property and Indigenous Sovereignty: Landscape and Literature in Australia,' *Halfway House, the Poetics of Australian Spaces,* ed, Jennifer Rutherford and Barbara Holloway, (Western Australia; UWA Publishing, 2010), 159.

49 Gavin Malone, 'Phases of Aboriginal Inclusion in the Public Space in Adelaide South Australia since Colonisation,' (PhD diss., Flinders University, 2012), 2.

nephew for burial. Each time he stopped, he was overcome with sadness and his tears formed the fresh water springs that string the coastline.[50] Four of the springs at Red Ochre, Pt. Willunga, Wirrina and Second Valley formed points of intersection between my own personal history and my evolving awareness of Aboriginal habitation. Somehow the knowledge of how to get fresh water on the beach at Port Willunga had been transferred to my family but it was a long time before I began to understand the significance of the magical welling between high and low tide.

As I traced my parallel costal journey to that of Tjirbruke, save for the place-names of Willunga and Aldinga, the evidence of Aboriginal habitation was less tangible than the solid stone and slate colonial architecture hunkered down throughout the town of Willunga or the huge dam wall holding back water that lay over the top of the sunken town of Myponga. However, this apparent slightness of materiality concerning Aboriginal inhabitation belies the complexity, inventiveness and durability of Aboriginal arrangements about the maintenance of myths, subtle kinship bonds and an abiding relationship to country.[51] The question arose as to how these discordant notes of 'incumbent and incursive culture'[52] could be heard, how it was that postcolonial experience coloured my understanding of landscape.

In some strange way I had come to understand that this landscape was also a 'ruin' in the sense that it consisted of the fragmentary remains of the process of colonial incursion.

7 Unsettlement

As the theorists Ken Gelder and Jane Jacobs explain, postcoloniality presents a conundrum for contemporary Australians who may simultaneously sense themselves as 'not being implicated in earlier processes of colonization,'[53] while at the same time having a collective sense of guilt about the procedures involved. This may go some way to explain a childhood sense of unease I had about a rift between the Aboriginal place names and the landscape they

50 Ibid., 2.

51 Latour, *An Inquiry into Modes of Existence*, 231.

52 Ross Gibson 'Shot Reverse Shot: Images and Testimonials Regarding the South Australian Colonial Experience Just After the Frontier,' (masterclass), University of South Australia, August 22, 2014, np.

53 Ken Gelder and Jane M. Jacobs, *Uncanny Australia Sacredness and Identity in a Postcolonial Nation*, (Melbourne: Melbourne University Press, 1998), 24.

described, Aldinga meaning place of wild ducks, Willunga denoting place of tall green trees and Myponga which meant place of tall cliffs. In each case the landscape described in the meaning did not fit. This presented disturbing unanswered questions about settlement and displacement. Gelder and Jacobs articulate 'postcoloniality as a contemporary moment, where one remains within the structures of colonialism even as one is some how located beyond … or "after" them.'[54] And so as I walked through this area I recalled the feeling of being surrounded by patterns of absence. Gelder and Jacobs, quoting Gibson, explain that from the position of incursive settler culture 'one can imagine being "in place" in Australia only through the realization that one is also "out of place."'[55]

And so the troubling evidence of postcolonial convocation which had changed the landscape through changing land use can be seen in the work of Tasmanian artist Raymond Arnold who investigates the immersive experience of landscape around the Queenstown area of Tasmania. Here geology has been laid bare by the former act of mining. This area also borders the Tasmanian Wilderness where the competing interests of mining, forestry and conservationists are in a constant state of rumbling debate.[56] Arnold describes his connection to the rocky environment as a form of psycho geography,[57] which tallies with the idea that landscape is also an internalised experience as well as a physical experience.

8 Place and Space

In the beginning of my research the journey from Mylor to Second Valley existed as an *ekphrasis,* the unknown areas occurring in my imagination as spaces rather than places. As the cultural geographer Tim Creswell explains, places are often associated with where we live or where we lived when we were young. They can be fixed to a geographic co-ordinate, have qualities of materiality and, more importantly, are made meaningful through human attachment. People imbue places with significance.[58] Spaces however have more abstract

54 Ibid., 24.

55 Ibid., 28.

56 Raymond Arnold, *Landscape art research Queenstown*, viewed 14 December 2014, http://landscapeartresearchqueenstown.wordpress.com.

57 Raymond Arnold, *Julian Cooper,* (Tasmania: Landscape Art Research Queenstown), 2012, 3.

58 Cresswell, Tim, *Place a Short Introduction,* (Melbourne: Blackwell Publishing, 2008), 7.

qualities, such as 'outer-space or the spaces of geometry.'[59] And so these qualities give space a sense of potentiality.

In Donald Meinig's essay 'The Beholding Eye Ten Versions of the Same Scene' (1979), differing codified views are used to frame the discussion of what we see in landscape. The landscape artist, he believes, articulates a distillation of the essence of landscape whilst delving into his or her own mysterious world. This view of landscape, Meinig ascertains, 'lies beyond science, holding meanings which link us as individual ... psyches to an ineffable and infinite world.'[60] Visual representation of the landscape and the bestowal of place names, therefore, are ways of anchoring space and attaching broader cultural significance to somewhere that becomes a place.

High above the beach near Wirrina was the place where I chose to end the long walk. Eventually on day six, after walking along an old coach road above the shore, I stopped. Tucked into the long grasses high above the shoreline, I watched people and dogs coming and going below me. To my left I looked out on where the folded hills met the sea. The rounded knolls gave way to dark scattered rocks, which receded into the distance. I could see the white water chopping at the long seawall jutting out of Wirrina Cove. And to my right I could see the extended line of Norfolk Island pines flanking the coast road growing smaller. A thin strip of white sand across the rough bay showed the beach at Carrikalinga. Far off in the misty distance the low-slung Adelaide Hills where I had started stretched away with Mt Lofty lying just proud of the ridge in the middle.

9 The Act of Place Making

In September of 2014, I presented aspects of my research relating to this long walk at the '4th Global Conference on Time Space and the Body' at Mansfield College, Oxford. The interdisciplinary nature of the conference provoked discussion by delegates from around the world concerning the ability different spaces have of shaping the body culminating in the publication of this anthology. What surprised me about the ideas exchanged was how exotic Australia still seemed to most people, resonating with Nycole Prowse's analysis in this volume of Pablo Armenillo's depiction of Australia as an 'ob-scene' space and a

---

59 Ibid., 8.

60 Donald W. Meinig, 'The Beholding Eye: Ten Versions of the Same Scene.' *The Interpretation of Ordinary Landscapes: Geographical Essays*, (New York: Oxford University Press 1979), 12.

region 'down under.'[61] The delegates were intrigued by the landscape and scale of distance, how for them it existed as an imaginative space almost as if it were a fictional alternative world full of strange reversals, somehow familiar but 'transposed into unfamiliar space.'[62] Whereas for me, it was the place I come from and the artworks created out of this relationship to place also referred to the unsettling ideas of colonial intrusion and the burden of history. And so the long walk had ended in the creation of artwork as an act of placemaking. In my studio, away from the sites of walking, the mood of these experiences were ruminated and layered into paintings that recorded the capricious conditions of the outside world, layers that were reminiscent of the historical layers that I had uncovered in the landscape.

Place is, therefore, most importantly an internal query that sits at the heart of questions of self. The critical theorist Timothy Morton, quoting Arne Naess, explains that the self is always inseparably part of the environment, existing as a relational junction within the total field of nature. The process of self identification therefore grows out of this relationship juncture.[63] Place is, moreover, a complex set of ideas which can include the descriptive particularities of a geographic area, the processes of social construction or a phenomenological interest in how humans exist 'in place.'[64] These overlapping ideas are neither bounded nor static. For me, it was the ideographic tethering process of painting and also in traversing the unknown areas between the places that had brought together incongruent elements attached to the material landscape, making them places and no longer spaces.

This investigation of landscape was a generative investigation where the aperture of the system moved back and forth between walking, research and the creation of artwork.[65] What I also found when I undertook the long walk from Mylor to Second Valley over a period of nine months was a tract of land underpinned by patterns of absence and the areas I lived as a child lying in wait to release their dormant memories as I passed. Instead of using a map I had followed the desire lines inscribed by animals and people, a landscape that was also engraved by the movement of settlement. I traversed the physical landscape of the present where traces of the past broke through a thin historical fabric.

61 Nycole Prowse, 'Possibilities from the Peripheries: Time Space and the Body in the Australian Drug Writing of Helen Garner and Luke Davies,' in this volume.

62 Turcotte, 'Australian Gothic,' 1.

63 Timothy Morton 'John Clare's Dark Ecology,' *Studies in Romanticism*, 47–2 (2008), 186.

64 Cresswell, *Place a Short Introduction*, 51.

65 Ross Gibson, 'Shot reverse shot,' np.

For Australian contemporary artists tension arises, therefore, between the aesthetic contemplation of landscape and the transformation of landscape through postcolonial convocation. The imaginative space of the walk I undertook was instilled with this discord between colonial and gothic landscape. Whilst early views of Australia created an impression of Arcadia, Australian gothic literature conversely revealed a dark and desolate space, which alluded to guilt and anxiety and conjoined landscape with otherness. This atavistic dissonance could also be heard in the use of European place names by early explorers, which signified a bucolic erasure of Aboriginal history. On the other hand, the use of Aboriginal place names in the area where I grew up served to highlight patterns of absence and a sense of postcolonial unsettlement. In this contested space a complex alignment of whispering cultural influences tugged and pulled at one another. This area was, moreover, a lived space and like other contemporary artists who make sense of the landscape through the immersive act of walking I was able to sense and intuit these forces through an incremental perceptual engagement. Making sense of the landscape area between Mylor and Second Valley was, therefore, a form of cognitive mapping as well as an act of placemaking.

Bibliography

Arnold, Raymond. *Julian Cooper.* Tasmania: Landscape Art Research Queenstown, 2012.

Bonython, Warren. 'History of the Heysen Trail.' Viewed February 11 2015. http://www.heysentrail.asn.au.

Carter, Paul. *The Road to Botany Bay.* London: Faber and Faber, 1987.

Cresswell, Tim. *Place a Short Introduction.* Melbourne: Blackwell Publishing, 2008.

Gelder, Ken, and Jacobs, Jane M. *Uncanny Australia Sacredness and Identity in a Postcolonial Nation.* Melbourne: Melbourne University Press, 1998.

Gibson, Ross. 'Shot reverse shot: images and testimonials regarding the South Australian colonial experience just after the frontier, (masterclass),' University of South Australia, August 22, 2014.

Gibson, Ross. 'The Known World,' *Text: Symposium: Creative and Practice-led Research,* (2010): 1–11.

Gilloch, Graeme. *Myth & Metropolis Walter Benjamin and the City.* Oxford: Polity Press, 1996.

Gregg, Simon. *New Romantics Darkness and Light in Australian Art.* Melbourne: Australian Scholarly Publishing, 2011.

Guthrie, Camille. 'Writing Poetry About Art,' *Poetry Foundation*. Viewed May 20, 2014. http://www.poetryfoundation.org/.

Hill, Shona, 'Imprinted on the Body: Changing Sensibilities in Time and Space,' in this volume, 2018.

Horne, Julia. *The Pursuit of Wonder, How Australia's Landscape was Explored, Nature Discovered and Tourism Unleashed.* Melbourne: Melbourne University Publishing, 2005.

Hughes, Robert. *The Fatal Shore.* London: Pan Books, 1988.

Kelly, Mimi. *Stranger in a Southern Land.* Sydney: A-M Gallery, 2014.

Kosma, Anthi. 'Spacing by Trace: an Event of an Out of Place Experience,' in this volume, 2018.

Latour, Bruno. *An Inquiry into Modes of Existence, and Anthropology of the Moderns.* Translated by Catherine Porter, Cambridge, Massachusetts: Harvard University Press, 2013.

Linn, Rob. 'The Discovery and Settlement of the Fleurieu Peninsula and the Angas/Bremer Region 1802–1861,' *The Journal of the Historical Society of South Australia.* (1986): 51–66.

Lippard, Lucy. *The Lure of the Local Senses of Place in a Multicenterd Society.* New York: The New Press, 1997.

Macfarlane, Robert. *The Old Ways.* New York: Viking, 2012.

Malone, Gavin. 'Phases of Aboriginal Inclusion in the Public Space in Adelaide South Australia since Colonisation.' PhD Dissertation, Flinders University, 2012.

Malouf, David. 'Peripecia,' *Mandy Martin: Peripecia, the Salvator Rosa Series.* The Australian National University: Canberra, 2002.

Mathews, Eric. *Philosophy of Merleau-Ponty.* Durham: Acumen Publishing, 2002. Monro, Clementine. 'Perpetual Transition: A communication between photography and absence within the ruin,' in this volume.

Meinig, Donald, W. 'The Beholding Eye: Ten Versions of the Same Scene.' *The Interpretation of Ordinary Landscapes: Geographical Essays.* New York: Oxford University Press, (1979): 33–48.

Orr, Jill. 'Heat: Art, People and Climate Change,' *Transdiscourse 1: Mediated Environments.* Edited by Andrea Gleiner, Angelika Hilbeck, Jill Scott, Springer: New York (2011): 105–116.

Plate, Cassie. German Romanticism, British Property and Indigenous Sovereignty: Landscape and Literature in Australia," *Halfway House, the Poetics of Australian Spaces.* Edited by Jennifer Rutherford and Barbara Holloway, Western Australia; UWA Publishing (2010): 159–169.

Rolls, Eric. *A Million Wild Acres.* Melbourne: Penguin, 1984.

Smith, Russell. 'The Spatial Poetics of Computer Games and the Mapping of Australian Space,' *Halfway House the Poetics of Australian Spaces.* Edited by Rutherford,

Jennifer and Holloway, Barbara Western Australia: University of Western Australia (2010): 345–365.

Steel, Kathleen. 'Fear and loathing in the Australian bush: gothic landscapes in Bush Studies and Picnic at hanging Rock,' *Colloquy: Text Theory Critique.* Melbourne: Monash University, (2010): 33–56.

Turcotte, Gerry, 'Australian Gothic,' *The Handbook to Gothic Literature.* Edited by Roberts, Mulvey. Basingstoke: Macmillan, (1998): 10–19.

CHAPTER 11

Spacing by Trace: Four Events of an Out of Place Experience

Anthi Kosma

Abstract

This chapter attempts to compose a collage between memories of events from a personal experience of the author and the references that other authors contribute to describe drawing as action, a 'gesture' through which the body externalises itself. Drawing is described as an action of coming-into-presence, which means presence before signification. It is also defined as a non-representative art where the form is not the traced figures but the body as a form by itself, the body as generator of space. Four short narratives are used in a phenomenological approximation to describe body from an inside point of view where spacing by trace is experienced. The act of making traces appears out of intimacy and exteriorises – through a dynamic and diagrammatic kind of writing – aspects of the abyss of oneself in a process where 'every time is a singular time.'[1] Drawings are ruins, testimonies of these moments, moulds of the gestures that traced them. The experience of drawing is also 'an interior matter,' an experience that disappears when the action stops and a performing art begins, which is always accompanied by images in 'a phantasmatic dance.'[2] If contemporary drawing is defined by Alain Badiou as a 'description without a place,'[3] this chapter looks forward by describing events where space is presented in its formation as an out of place description. 'Out of place' because drawing is not perceived or not concerned as a given, available or formed form. On the contrary, it is the gift, invention, uprising or the birth of form.

1 Introduction-Events

The description is an 'artificial thing that exists.' Artificiality. Drawing is something which is composed. It is the question of technology. Today the

1 Jean-Luc Nancy, *The Pleasure in Drawing* (New York: Fordham University Press, 2013), 21.
2 Giorgio Agamben, *Ninfas* [*Nymphs*] (Valencia: Pre-Textos, 2010), 8.
3 Alain Badiou, 'Drawing.' Lacanian ink 28 (2006): 42–48. September 12, 2011. v viewed on 25 January 2015, http://lacan.com/symptom12/?p=65.

> background can be a screen, and not a piece of paper, and the marks can be the visible projection of immaterial numbers. [...] that a Drawing is a work of art has a precise meaning. It is a description without place which creates a sort of artificial world. This world does not obey the common law of separation between real being and appearances. In this world, or at least in some points of this world, there is no difference between 'to be' and 'to exist'; or between 'to be' and 'to seem,' to 'appear.'[4]

The chapter attempts to compose a collage between characteristic events from a personal experience of the author and the references that other authors contribute to describe the action of tracing and precisely in that of drawing and the interaction between body and its ambience at the nascence of the figures. Personal experiences in drawing and bibliographical references form a fictional story based on real facts. In this story characteristic aspects of the experience of drawing in relation to the body in the experience of spacing are described. The experience in drawing that this chapter collects is the one developed by the author at the context of the department of D.I.G.A. at the E.T.S.A. (School of Architecture) of Madrid,[5] the same that formed the subject of a practice-led investigation about the action of drawing[6] and the object of this chapter.

In addition reference to the authors that participate in this volume are used in order to link and enrich the description for the body while it is tracing. Links between this chapter and that of others aim to enforce this net (of authors) and to facilitate a transversal reading of this collection of texts about 'time, space and the body.' Linking this text about the experience of spacing through the traces of drawing with other cases and approaches about the body offers a broader view both of the experience of spacing and the body in general and of the action of drawing and tracing itself in particular. Moreover, through this net of texts and the links between them the action of drawing is presented as a kind of writing, an act of exteriorisation in relation to other cases. Other cases

---

4 Alain Badiou, 'Drawing.' Lacanian ink 28 (2006): 42–48. September 12, 2011. v viewed on 25 January 2015, http://lacan.com/symptom12/?p=65.

5 Department of Graphic Design in the Architectural School of Madrid. In particular at the curses of 'Advanced Drawing and Graphic Interpretation' (DAII) I y II' and 'Life drawing,' both taught by professor Antonio Verd Herrero and the active participation and the work of the investigation and Innovation Educational Group 'Hypermedia,' coordinated by professor Javier Seguí, during the period starting the year 2007 until the present (2014). This experience was also used as fundamental part of my PhD thesis and of other experiences based on drawing.

6 Anthi Kosma, 'Proyectar Dibujando: Una Aproximación Fenomenológica al Estado Naciente del Proyecto. Un estudio Entre dos Culturas Arquitectónicas.' PhD diss., Arquitectura, 2014.

manage to present drawing not as an action a language for specialists but far from it, as a common language and a corporeal way to experience the space, to feel the experience of spacing.

This 'out of place'[7] description consists of four short events. Each event presents a different point of view on the issue. These points of view are incomplete as units but investigated as a group they form a draft, a kind of cartography of the different aspects of the act of drawing. The first event shows how drawing as an action without finality, as non-representation, provokes estrangement and how the body, through this process, gives birth to figures, opens and exteriorises itself leaving fragments of its existence over a surface. In the second event, drawings are presented as objects that are the remaining of the movements of a corporeal dance where the hands are tracing. These remains can be 'read' and interpreted by the action of drawing and can be referred and connected to this unknown and unfamiliar body. During the third event the experience of spacing during the birth of form is described from an inner point of view. 'Drawing is the opening of form'[8] where form is the traced images but also the form of the body. Finally, the fourth event completes these brief descriptions explaining how this opening, this dance over a surface is complicated, impossible to be described without making reference to the images and to the figures that arise and appear during the action of tracing. Images, however, are not only the traced images but also images of the body itself while it is moving.

Bibliographical references about drawing in combination with empirical content facilitate the description of the experience of drawing and simultaneously that of the body from an inside point of view, as an inner experience, an exploration of space and its limits by itself.

2 First Event: A Movement without Purpose

> Sit at a table with pen and paper; put yourself in a 'receptive' frame of mind, and start ~~writing~~ drawing. Continue ~~writing~~ drawing without thinking about what is appearing beneath your pen. ~~Write~~ Draw as fast as you can. If, for some reason, the flow stops, leave a space and immediately begin again by ~~writing~~ drawing down the first letter of the next

7 Alain Badiou, 'Drawing.' Lacanian ink 28 (2006): 42–48. September 12, 2011. viewed on 25 January 2015, http://lacan.com/symptom12/?p=65.

8 Jean-Luc Nancy. *The Pleasure in Drawing* (New York: Fordham University Press, 2013), 1.

> sentence. [...] What is crucial is the unpremeditated free association that creates the basic text.[9]

There have been many, almost innumerable, beginnings since the first time I experienced this peculiar way of drawing. The automatic drawing, based on the principles of the automatic writing of the surrealists, was the introduction to a pedagogy on drawing completely different than the artistic education on drawing I had had by this time. Drawing is perceived as 'this kind of writing,' according to Roland Barthes, that 'traverses the aesthetic stereotype' converting the surface in an 'allusive field of writing.'[10] Drawing or writing is the act of trace over a surface in a process where the experience of space is presented in its formation and not as a mere object or as a reflection about a finished object.

The case of automatic drawing with the spontaneous generation of forms facilitates this 'out of place description' as an example that permits the observation of the body in action, where the body as a gesture exteriorises itself. Jean-Luc Nancy's words about the pleasure of drawing describe this state of the body well.

> Drawing is represented, experienced, and experimented with as a compulsion, like the effect of an irresistible impetus. [...] open to joy or sensual pleasure [...] a form of pleasure, but a pleasure that is not satisfied by grasping an object: it satisfies itself in transporting a subject outside itself.[11]

Action is used in this chapter in a way that aspires to be related to the way that is described by the words of Hannah Arendt in her text about freedom and art work:

> We deal here not with the liberum arbitrium, a freedom of choice that arbitrates and decides between two given things, one good and one evil, and whose choice is predetermined by motive [...] That is, the freedom to call something into being which did not exist before, which was not

9 Alastair Brotchie and Mel Gooding, eds. *Surrealist Games* (Boston: Shambhala Publications, 1993), 45.

10 Roland Barthes, 'Cy Twombly: Works on Paper' in *The Responsibility of Forms, Critical Essays on Music, Art and Representation* (New York: Hill and Wang, 1985).

11 Jean-Luc Nancy, *The Pleasure in Drawing* (New York: Fordham University Press, 2013), 56–58.

> given, not even as an object of cognition or imagination, and which therefore, strictly speaking, could not be known.[12]

'No drawing is known, we draw to know.'[13] This state of the body in action also requires a state of flow where flow

> Involves new procedures or requires an unexpected variation. [...] The flow state tends to occur when a person's skills are fully involved in overcoming a challenge that he [or she] may face.[14]

It must have been the pleasure of this kind of dance of free movements and a feeling of estrangement each time a new itinerary and a new choreography were traced that probably caused this kind of engagement I experienced. This way of triggering a process gives not only the process without a specific ending but also without explanation to this way of drawing and is being used in this chapter as the object of the study. In particular, in this context and also in this chapter, drawing is perceived not as an object, neither as representation[15] nor as a technique, a work or a discipline exclusively preparatory[16] but as action, exploration, way of searching, experimental medium.'[17] In this process drawing is presented not as the result of an external motivation and reference but it is imminent to its proper action, a gesture, an 'echography' between the myth and the project.[18]

---

12 Hannah Arendt, *Between past and future: Eight exercises in political thought* (New York: Penguin, 1968), 151.

13 Juan Luís Moraza, *Oteiza Laboratorio de Papeles: El reverso del dibujo: Notas sobre la noción de proyecto en 840 esbozos de Oteiza* [Oteiza Laboratory Papers : The reverse of the drawing : Notes on the notion of project outlines Oteiza 840.] (Navarra: Foundación Museo Oteiza, 2006), 146.

14 Mihaly Csikszentmihalyi, *Aprender a Fluir*, [Learn to flow] (Barcelona: Kairos, 1998), 43.

15 Javier Seguí de la Riva, *Dibujar, Proyectar* (IX)*: El grado cero de la arquitectura* (Madrid: Instituto Juan de Herrera de la escuela de Arquitectura de Madrid), 2003.

16 Steve Garner, *Writing on Drawing, Essays on Drawing Practice and Research* (England: NSEAD, 2008), 24.

17 Irish Museum of Modern Art, 'What is drawing?' February 2013, viewed on 9 October 2014, http://www.imma.ie/en/downloads/whatisdrawing2013.pdf.

18 Juan Luís Moraza, *Oteiza Laboratorio de Papeles: El reverso del dibujo: Notas sobre la noción de proyecto en 840 esbozos de Oteiza* [Oteiza Laboratory Papers : The reverse of the drawing : Notes on the notion of project outlines Oteiza 840.] (Navarra: Foundación Museo Oteiza, 2006), 146.

When the body and the hands leave their traces without intention in 'a compulsion, like the effect of an irresistible impetus,'[19] in a process that no longer is looking forward to represent finished well known objects according to some rules, many characteristics of the body and its reactions can be noticed. Drawing as action and gesture and drawings as objects-echographies of a body in exposure, facilitate the observation of fragments of the body 'outside itself'.

This dislocation of the subject outside itself is where the experience of spacing by trace, by drawing, is felt. Due to this point of view in this chapter space is not perceived as a mere object nor is in presence but it becomes apparent in this gesture of exteriorisation, in this exposure of oneself during the appearance of the traced figures 'out of place.' Drawing is also described as an 'action of coming-into-presence [which] means presence before signification'[20] and also as a non-representative art where the form is not the traced figures but the body as a form by itself, the body as generator of space.

> Drawing participates in a semantic field where act and force [puissance] are combined, or where the sense of the act, the state, or the being that is in question cannot be detached entirely from a sense of gesture, movement, or becoming.[21]

In this process of tracing where 'each time [is] single and singular'[22] or when other kinds of reactions and unknown ways of exploration are being looked for, the body and its actions have an essential role. The routine, the repetition of drawing as craftsmanship, coexists in a challenging balance with the effort to search, break and go against the known, to provoke new forms and openings. The focus is on drawing as an action, not as a representation, as a way of exploration, a 'purposiveness without purpose.'[23] It is an action that takes place here and now, from an inside point of view of the person that traces, writes or draws – making it easier to refer to the production of spacing.

19 Irish Museum of Modern Art, *'What is drawing?'* February 2013, viewed on 25 January 2015, http://www.imma.ie/en/downloads/whatisdrawing2013.pdf.

20 Jean-Luc Nancy, *The Sense of the World*, trans. Jeffrey S. Librett (Minneapolis: University of Minnesota Press, 1997), 128.

21 Jean-Luc Nancy, *The Pleasure in Drawing* (New York: Fordham University Press, 2013), 21.

22 Ibid., 3.

23 Ibid., 93.

3 Second Event: Drawings, Body 'Moulds'

> We put the papers on the wall. And then, silence. It was the moment that we had to speak about these spots and interpret what we had been drawing. It wasn't so easy. We had to speak about the drawings and also about the relation between the traced figures and their author. Imagine the possible ways the author had followed to trace the image, search among the signs and 'revive' these possible 'battles.' In other words we had to invent a story about the state of the author while drawing and what these abstract signals would like to transmit.[24]

In drawing, just like in writing, after the act of tracing comes the act of reading. At first, 'reading' these non-significant signs seems impossible. Signs are 'silent' and at first sight, without sense or significant connotation. Words as significant phonemes come to break the hermetic silence of the figures.[25]

Clementine Monro refers to this silence, in the case of architectural ruins, in this volume.[26] Experiencing the architectural ruins in their actual state, away from the way they were built or used, is presented as a response of the body in interaction with them. 'Absence has the capacity to embody the fullness of presence' states Monro and dedicates her article to the 'observation of absence': 'The ruin in contrast often gives us nothing of artefacts and facts; it is instead an empty space, yet through absence and the embodiment of presence, it is full.'[27] This response of the body and feeling of fulfilment in its interaction with the art work comes from an *aesthetical* point of view. The experience of spacing by the trace of drawing is approximated by a *poetical* point of view.

For this 'out of place description' of the action of drawing, absence is referred to the body during the process of *poesis*. Drawings evoke the absence of the body and its gestures during the formation and the appearance of the traced images as it exteriorised itself. Drawings as remains of this action are the testimonies that can help us to 'regress' at the moment of the formation of the art work and describe the dislocation of oneself while spacing by trace is experienced. Monro suggests that '[a]bsence requires context'[28] and in the case of drawing as it is described in these lines the context is the body itself, the

24 Memories of the author drawing classes at ETSAM, UPM, 2008–2012.

25 Antonio Verd Herrero, notes from DAEIG lesson at ETSAM, UPM, 2010.

26 Monro Clementine, 'Perpetual Transition: A Communication Between Photography and Absence Within the Spatial Experience of the Ruin,' in this volume.

27 Monro, in this volume.

28 Monro, in this volume.

body during the action of tracing. Jacques Rancière alludes to this in *Aisthesis: Scenes from the Aesthetic Regime of Art:*

> A work exists, as a self-sufficient unity, to the extent that the potentiality of an open whole is expressed within it, a whole that exceeds all organic totality. Strictly speaking, there are no forms. There are only attitudes, unities formed by multiple encounters of bodies with light and other bodies. These attitudes could also be called surfaces. For surfaces are something entirely different than the combinations of lines; they are the very reality of everything that we perceive and express. [...] Dramatic action and plastic surface can be reduced to the same reality: the modification of this large, vibrant surface, excited and modified by a unique force called Life. One should not misunderstand the apparent banality of the word 'life'... [and see] the resolution of the perennial tension between three terms: 'body,' 'life,' and 'action.'[29]

These attitudes were what we investigated at the sessions of drawing when we put the papers on the wall trying to imagine possible ways that the author of a traced image had followed. Attitudes, gestures, ways of moving, dispositions, ways that the body might have transformed itself were figured out in order to prove the unity between body and action.

John Dewey's notion of 'regress to the experience,'[30] helps explain the moment of creation and the conditions and the ways that the author produced an artwork – the mental effort that offers the chance to fabricate a possible story about the way a body, in certain conditions and environment, made it. 'Regress' to the action requires a certain position of the observer and it is an indispensable condition in order to understand the art work. In the case of drawing as an object, traces remain as residues of an archaeological excavation, testimonies of a body in a certain state, a certain moment and way of movement. Juan Luís Moraza states:

> There are no signs, a trace cannot lie, but psychic onomatopoeias, accomplishments, knotting, in their curbs, in their fictions, their own pretences reveal immediately, as oscilloscope or echography [...] just observe them closely enough to find out this immediate quality unrelated

29 Jacques Rancière, Aisthesis: Scenes from the Aesthetic Regime of Art (London: Verso, 2013), 163.

30 John Dewey, *El Arte Como Experiencia*, Trad. Jordi Clara monte (Barcelona: Paidos, 2008), 12.

> to any disquisition, to any mediation.[…] in any case it's a double expansion, a double folding: towards itself, the trace intensifies the experience – generates fields of internal complexity; toward outside expands the body far away from the presence and the perception: material of memory marks.[31]

After the action traces stay there as testimonies of a back side of a gesture, memories of a moment, impossible to represent and evoke. They are perceived also as 'moulds' of a body in action in its opening, materialities that reveal their positive also in constant alteration. 'The contra mold [sic] of the image'[32] as François Jullien states:

> All these traces are closed to this graphic contra-mold [sic] which is the union of all of them […] an inner drawing […] the trace that is the trail of the body, its longing and hesitation that not only transmits ideas and projects, but at the more real of the desire of a subject.[33]

Only evolved by drawings, by traces, this peculiar and diffused body is presented, it can be felt and touched where, as Javier Seguí de la Riva describes, 'each artist acts as a silkworm that generates its cocoon secreting its thread, while he stirs in the space in a continuous, repetitive, self reference ovoid dance.'[34]

This reversed approximation from the drawings as ruins to the drawing as the action while the body is tracing, from an outside point of view of the drawing to an inside point of view is a process of unmaking and at the same time a process of construction of a possible narrative. In other words this event, dedicated to the interpretation of the art work of drawings as gestural ruins, is a fabricated story for the corporeal experience of spacing by trace which make us perceive drawings as objects inseparable from the body that traced them.

I will not forget the day when in one session of drawing they told us 'Take the blank page in front of you and crumple it.' And then they asked 'Is this a

31 Juan Luís Moraza, *Oteiza Laboratorio de Papeles: El reverso del dibujo: Notas sobre la noción de proyecto en 840 esbozos de Oteiza* [Oteiza Laboratory Papers : The reverse of the drawing : Notes on the notion of project outlines Oteiza 840.] (Navarra: Foundación Museo Oteiza, 2006), 145–160.

32 François Jullien, *La Gran Imagen no Tiene Forma* [The big image has no form]. (Madrid: Alpha Decay, 2008), 349.

33 Ibid., 349.

34 Javier Seguí de la Riva, *Movimiento y Reposo* Notes. Viewed on 25 January 2015 http://javierseguidelariva.net/Trabajos%20en%20curso/Excerpta/09.En%20movimiento%20y%20en%20reposo/A.Movimiento%20y%20reposo.pdf.

drawing?' 'Are you able to speak about the person who traced it?' Maybe since this moment I began to realise that the body is leaving its traces not only in the surface of a paper or a screen but all over the material world in different formats, scales and within different mediums. In other words, in a certain way, human traces are all over, were 'remains' related to this gesture of exteriorisation of the body.

4 Third Event: An Interior Matter

> A writer doesn't touch by grasping, by taking in hand but touches by way of addressing himself, sending himself to the touch of something outside, hidden, displaced, spaced. […] the body is the stranger 'out there' (the place of all strange things) because it is here. Here, in the 'there' of the here, the body opens, cuts, displaces the out – 'there.'[35]

In this description, as observers we are located inside the action, from an inside point of view, from the point of view of this body that ignores its limits and its external form, which is not the skin. This inner point of view of the body is for instance almost an 'inverted' description of the body than that which Shona Hill presents in her article 'Imprinted on the Body: Changing Sensibilities in Time and Space.'[36] In Hill's article, the body is presented from an external point of view. There the perception and knowledge of our body is coming from outside, from the representations that media offered to us. Hill claims that by examining the body in diverse media forms there is 'a manifestation of a social reality inside the picture' of the body.[37] This approximation reveals how references from outside manage to manipulate, enable or disable sensibilities toward the human body and our perception of it.

However, what is presented in this chapter is a different approximation. The body is perceived from an inner point of view while it is in action where its perception cannot be defined: '[f]rom inside, in the miniaturization, the abysm of the fathomless center opens.'[38] For Javier Seguí de la Riva the experience of drawing reaches a point where the body is converted into a 'being drawing':

---

35 Jean-Luc Nancy, *Corpus* (New York: Fordham Univ Press, 2008), 17–18.

36 Shona Hill, 'Imprinted on the Body: Changing Sensibilities in Time and Space,' paper presented at the '4th Global Conference on Time, Space and the Body.'

37 Shona Hill, 'Imprinted on the Body: Changing Sensibilities in Time and Space,' paper presented at the '4th Global Conference on Time, Space and the Body.'

38 Javier Seguí de la Riva, *Ser Dibujo.* (Madrid: Mairea libros, 2010), 73.

> Being drawing is to see everything as drawn and feel the drawing (the making) as the essence of being alive plowing the delineation. Being drawing is to feel that identity is the accumulated sediment compaction, the result of a life that has been spent moving, making figures and traces. Being drawing is being seen as a product of a fact, as the work of making works, as an erased figure.[39]

The differences in approach are significant – Hill's 'imprinted body' is described from the outside and the body as a 'being drawing' is described from an inside point of view, a body while it is tracing. During the action, as Hannah Arendt defines it, there is no longer consciousness of our acts and, therefore, the perception of the body is not possible. Only after the action, when (in this case) the process of drawing is completed, consciousness of the body and its actions is possible. Descriptions of this inner point of view of the body are narratives based on memories of the body during the action as mediation in a gesture of exteriorisation. This does not mean that these descriptions cannot be affected by external references and criteria as in the case of the 'imprinted body' but these empirical descriptions dedicated to drawing and making lack of the external image of the body. The body is not an object with certain limits but an extension in constant movement and alteration: 'the body is immaterial. It is a drawing, it is an outline, it is an idea.'[40]

And in another text Jean-Luc Nancy acclaims:

> To begin with, I have to be in exteriority in order to touch myself. And what I touch remains on the outside. I am exposed to myself touching myself. And therefore – but this is the difficult point – the body is always outside, on the outside. It is from the outside. The body is always outside the intimacy of the body itself.[41]

According to this approximation, from an inner point of view, the body opens while it is writing and it opens towards the outside. As Jean-Luc Nancy states: '[d]rawing is the opening of form.'[42] For Nancy the opening is availability or inherent capacity. Drawing is opening of form because it is gesture, power, possibility and force (dynamics) that is performed and exercised (energy-*energeia*

---

39 Ibid., 121.

40 Jean-Luc Nancy, *58 Indicios Sobre el Cuerpo. Extensión del Alma* (Buenos Aires: La Cebra, 2010), 14.

41 Jean-Luc Nancy, *Corpus* (New York: Fordham Univ Press, 2008), 128–129.

42 Jean-Luc Nancy. *The Pleasure in Drawing* (New York: Fordham University Press, 2013), 1.

in Greek). It is action and not a mere artefact and for the description in this article the form that opens is the body, this unknown agent, the corpus 'as extension of the soul'[43] that exteriorises itself while tracing moves it.

We only manage to capture fragments of this body, never its totality. The body in its opening cannot be possessed and only through this opening is it possible to feel some of its aspects. It appears and disappears straight after the action and its exposition. Alain Badiou comments:

> Appearing is not like a formal body of being. [...] I can propose a short definition of Drawing: There is a Drawing when some trace without place creates as its place an empty surface. [...] There is thus no place for the pure spacing presupposed by every place.[44]

Maybe this definition helps us describe how space in the action of drawing can be considered more as spacing, where space is being experienced as an extension of the body and as an event of being with oneself and 'with others' in action through making by tracing.

In this chapter we focus on drawing as opening to oneself. Drawing in this description is self-referential. It always appears in relation to the movements, the state of the body that gives birth to the appearance of these traces. Drawing exists but does not look to anything else. It is an action or an object that has no relation with external references.

> Self consents to an impulse and not a given, to a birth and not a figure, it consents to an outside, to an other, to an alterity or alteration that, before all beginnings, will have mobilized the gesture without pre/scribing it by tracing out a path.[45]

The body by making-tracing opens towards the outside. Tracing, in this case, is the way to feel its limits, to be felt. It is a sense that disappears when the action stops. It can be said that this 'space temporal' approximation of the being, looking to grasp the sensation of the body from an inside point of view in a description after the action, is similar to the approximation of Nycole Prowse

43 Jean-Luc Nancy, *Corpus* (New York: Fordham Univ Press, 2008), 27.

44 Alain Badiou, 'Drawing' Lacanian ink 28 (2006): 42–48. September 12, 2011. viewed on 25 January 2015, http://lacan.com/symptom12/?p=65

45 Jean-Luc Nancy, *The Pleasure in Drawing* (New York: Fordham University Press, 2013), 33.

in this volume.[46] Prowse proposes that the literary discourse of drug writing enables a re-examination of the concepts of space, time and the body – where alternative corporeal possibilities and its limits are merged with the urban space, where portraying of 'the inner city is an urbanity.'[47] This notion of reconceptualising urban space in an empowering way, is also examined by Shelley Smith and Nicolai Steinø[48] in this volume.

5 Fourth Event: A Phantasmatic Dance

Spacing by trace is experienced in action, in movement. The movement of drawing as an action, a gesture of 'purposiveness without purpose' is considered as 'a phantasmatic dance' or better a 'dance of phantasms'[49] because this gesture of opening, of movement or touch is accompanied by images, by figures.

> How can an image charge itself with time? [Giorgio Agamben asks and finds the answer in the dancer] the dancer is not the body and its movement, but rather in the image as a 'Medusa's head,' as a pause that is not immobile but instead simultaneously charged with memory and dynamic energy.[50]

The dance of drawing takes place in a limited scene, a certain terrain background and foreground that is normally in the access-ratio of the hands. This dance, as it is presented here, is motivated by nothing other than the already traced or spontaneous gestures of the dancer-designer. Based on this metaphor between drawing as a movement and dance a connection with the chapter of another author in this edition is intended. Maybe at a first glance this volume's chapter by Smith and Steinø about the practice of parkour, known as the art of movement, has little to do with the action of drawing and the movement of the hands while they trace because of the different scale of its performance/movement. Smith and Steinø present the practice of parkour as a diverse way to use the public space in contemporary urban life, through the body and its

46 Nycole Prowse, 'Possibilities from the Peripheries: Time, Space and the Body in the Australian Drug Writing of Helen Garner and Luke Davies,' in this volume.

47 Ibid., 8.

48 Shelley Smith and Nicolai Steinø, 'Public Space Design between Alienation and Appropriation: The Case of Parkour,' in this volume.

49 Giorgio Agamben, *Ninfas* [*Nymphs*] (Valencia: Pre-Textos, 2010), 8.

50 Giorgio Agamben, *Ninfas* [*Nymphs*] (Valencia: Pre-Textos, 2010), 8.

unpredictable bodily gestures.[51] The 'traceur' is the person who takes part in the activity of parkour as the dancer of drawing but in a different scale and without mediums use:

> The art of freely combining an accumulated technical bodily knowledge. Technical knowledge involves the acquisition of all the gestural patterns that define scenic action (walking, running, climbing up and down, sliding, tumbling or tap dancing, slapping, using objects, catching and throwing a weight, archery or stabbing) [...][52]

Parkour 'dancers' negate the unilateral use of the urban space and with their performances, based on improvisation and spontaneity, manage to intervene, transform, break with certain attitudes and give another sense where 'architecture and design [are] considered as agents of control.'[53] The 'gymnastics of the impossible'[54] of the parkour dancers and the transformation that they offer with their theatrical performances present the body and the spatiality in a way that, similar to Prowse'[55] discussion of the narcoticising of the body and space of the city, calls for the abandonment of the traditional notion about space in the field of architecture. This change of focus from a fixed and controlled urban space to its use and transformation from a theatrical 'illogical' practice, might seem a generalisation but it can be considered similar to the way the notion of drawing is changing from a mere object to action.

> Drawing participates in a semantic field where act and force [puissance] are combined, or where the sense of the act, the state, or the being that is in question cannot be detached entirely from a sense of gesture, movement, or becoming.[56]

But dancers of the traced images have other kinds of complexities of space, time and the body than the ones of the parkour dancers. In reference to the 'traceur':

---

51 Smith and Steinø, in this volume.

52 Jacques Rancière, *Aisthesis: Scenes from the Aesthetic Regime of Art* (London: Verso, 2013), 90.

53 Smith and Steinø, in this volume.

54 Jacques Rancière, *Aisthesis: Scenes from the Aesthetic Regime of Art* (London: Verso, 2013),76.

55 Prowse, in this volume.

56 Jean-Luc Nancy, *The Pleasure in Drawing* (New York: Fordham University Press, 2013), 56–58.

> Images live within us. At this moment we each have an extensive visual world inside of us [...] We are living databases of images—collectors of images—and these images do not stop transforming and growing once they get inside us.[57]

So our body is inseparable from the images. Drawing new images appear and probably spacing is inseparable from the images too, existing through the action that reveals its presence. And certainly the body is inseparable from its proper image, the image of the body, the image of the body in action, when it traces towards the outside where drawing is the opening of the form, its form and the form of the body.

It is necessary to speak about formation, not of forms, action and not objects, for the conditions of the process of making, the effort the extended limits of the body, space as a corporeal experience, an experience always full of figures and images of 'phantasms.' This means that the experience of spacing is not a mere feeling of the subject while it is moving through traces but a dance of phantasms: the ones that appear with energy and others that are revealed through memory.

6 The Image of the Body, as an Epilogue

All events presented are only indications. This 'out of place description' is a brief mode to describe the body from an inner point of view, during the experience of tracing without purpose, through the remains of a non-significant way of writing, as a way to feel the space through the birth of the traced 'phantasms.' As a coincidence descriptions of the body, its gestures and the whole experience of spacing by trace cannot be separated from the traced figures, from their potency, their intimacy and strangeness, their phantasms and definitely not from the phantasm of the body, the 'image of the body.'

Within this chapter's discussion about the action of drawing and the experience of spacing, another kind of image has arisen. The image of the body while it is tracing is not the image of the drawings traced over a surface or a configured body but the image of this body in action, during the practice of tracing. This image of the body cannot be separate from the aspect of the body while it exteriorises itself through the action of tracing during the appearance of the traced images, it is a mnemonic image of the body experiencing its limits.

57 Giorgio Agamben, *Ninfas* [*Nymphs*] (Valencia: Pre-Textos, 2010), 13.

> Imagination is always considered to be the faculty of forming images. But it is rather the faculty of deforming the images offered by perception, of freeing ourselves from the immediate images; it is especially the faculty of changing images.[58]

In the case of spacing by trace, during the movement of drawing the changing images are the configured forms but also the forms of the body while it is moving, exteriorising itself through the practice of drawing. The image of the body is not a fixed image. Like the images of the dancer of Agamben, they are changing images, 'as a pause that is not immobile but instead simultaneously charged with memory and dynamic energy.'[59]

Bibliography

Agamben, Giorgio. *Ninfas* [*Nymphs*].Valencia: Pre-Textos, 2010.

Arendt, Hannah. *Between Past and Future: Eight Exercises in Political Thought*. New York: Penguin, 1968.

Bachelard, Gaston. *The Psychoanalysis of Fire*, trans. CM Ross Boston: 1964.

Badiou, Alain. 'Drawing.' *Lacanian Ink* 28 (2006): 42-48.

Barthes, Roland. 'Cy Twombly: Works on Paper' in *The Responsibility of Forms, Critical Essays on Music, Art and Representation*. Trans. Richard Horward. New York: Hill and Wang, 1985.

Brotchie, Alastair, and Mel Gooding, eds. *Surrealist Games*. Boston: Shambhala Publications, 1993.

Csikszentmihalyi, Mihaly. *Aprender a Fluir*. Barcelona: Editorial Kairós, 1998.

Dewey, John. *El Arte Como Experiencia*, Trad. Jordi Clara monte. Barcelona: Paidos, 2008.

Garner, Steve, *Writing on Drawing, Essays on Drawing Practice and Research*, UK: NSEAD, 2008.

Hill, Shona. 'Imprinted on the Body: Changing Sensibilities in Time and Space.' Time, Space & the Body, 4th Global Conference September 2014. Viewed on 25 January 2015 http://www.inter-disciplinary.net/probing-theboundaries/making-sense-of/time-space-and-the-body/project-archives/conference-programme-abstracts-and-papers-for-tsb4/session-3-concerning-bodies-in-space-and-time/

Jullien, Francois. *La gran imagen no tiene forma*. Madrid: Alpha Decay, 2008.

58 Gaston Bachelard, *The Psychoanalysis of Fire*, trans. CM Ross (Boston: 1964), 19.

59 Giorgio Agamben, *Ninfas* [*Nymphs*] (Valencia: Pre-Textos, 2010), 8.

Kosma, Anthi. 'Proyectar Dibujando: Una Aproximación Fenomenológica al Estado Naciente del Proyecto. Un estudio Entre dos Culturas Arquitectónicas.' PhD diss., Arquitectura, 2014.

Moraza, Jose Luís. *Oteiza Laboratorio de Papeles: El reverso del dibujo: Notas sobre la noción de proyecto en 840 esbozos de Oteiza*. Navarra: Foundación Museo Oteiza, 2006.

Nancy, Jean-Luc. *The Sense of the World*, trans. Jeffrey S. Librett (Minneapolis: University of Minnesota Press, 1997).

Nancy, Jean-Luc. *Corpus*. No. 67. New York: Fordham Univ Press, 2008.

Nancy, Jean-Luc. *The Pleasure in Drawing*. New York: Fordham Univ. Press, 2013.

Rancière, Jacques. *Aisthesis: Scenes from the Aesthetic Regime of Art.* London: Verso, 2013.

Seguí, Javier de la Riva. *Dibujar Proyectar* (IX): El grado cero de la arquitectura, Madrid: Instituto Juan de Herrera de la escuela de Arquitectura de Madrid, 2003.

Seguí de la Riva, Javier. *Ser Dibujo*. Madrid: Mairea libros, 2010.

Index

www.ingramcontent.com/pod-product-compliance
Lightning Source LLC
Chambersburg PA
CBHW071436230926
51846CB00001B/253
* 9 7 8 9 0 0 4 3 6 2 9 9 4 *